Civil Engineering Essentials

Civil Engineering Essentials

Sarah Woods

MURPHY & MOORE
www.murphy-moorepublishing.com

Published by Murphy & Moore Publishing
1 Rockefeller Plaza,
New York City, NY 10020, USA
www.murphy-moorepublishing.com

Civil Engineering Essentials
Sarah Woods

International Standard Book Number: 978-1-63987-107-0 (Hardback)

Cataloging-in-Publication Data

Civil engineering essentials / Sarah Woods.
p. cm.
Includes bibliographical references and index.
ISBN 978-1-63987-107-0
1. Civil engineering. 2. Engineering. I. Woods, Sarah.
TA145 .C58 2022
620--dc23

Contents

Preface

Civil Engineering is a primary branch of engineering which is concerned with the development and construction of buildings and physical structures. It employs the principles of physics, mathematics, hydrology and various other fields like industrial engineering, geography and geology to build structures such as dams, bridges, canals, etc. Civil engineering is further divided into various sub-disciplines such as structural engineering, coastal engineering, earthquake engineering, construction engineering and geotechnical engineering. Construction engineering involves site development and review of logistics costs. Earthquake engineering involves designing structures which can withstand earthquakes, while coastal engineering focuses on managing coastal area by keeping soil erosion and flooding in check. Other than construction work, civil engineering is applied to various other fields such as the aerospace industry to design space stations or the automotive industry to find the load carrying capacity of the chassis. Such selected concepts that redefine civil engineering have been presented in this book. It is a valuable compilation of topics, ranging from the basic to the most complex theories and principles in this field. Coherent flow of topics, student-friendly language and extensive use of examples make this book an invaluable source of knowledge.

A foreword of all chapters of the book is provided below:

Chapter 1 - The branch of engineering which deals with the professional practice of construction, design and maintenance of infrastructure such as bridges, roads, canals, pipelines, dams, airports, etc. is known as civil engineering. This is an introductory chapter which will introduce briefly all the significant aspects of civil engineering; **Chapter 2** - The determination of the physical properties of soil below the Earth's surface in a proposed installation site by using different methods is known as geotechnical investigation. This chapter encompasses the geotechnical investigations for dam, reservoir and tunnel and geological considerations for construction, etc. All the diverse principles of geotechnical investigations and geological considerations have been carefully analyzed in this chapter; **Chapter 3** - Construction refers to process and techniques employed in the assembly and erection of structures. Some of its important aspects of construction are construction engineering, classification of civil engineering projects, building construction, architectural engineering and structural engineering. The aim of this chapter is to explore the fundamentals of construction engineering; **Chapter 4** - There are various techniques used in the protection of a building against moisture. This chapter discusses some of them such as building envelope, damp proofing, interstitial condensation, vapor barrier, housewrap, basement waterproofing, etc. The topics elaborated in this chapter will help in gaining a better perspective about moisture protection in construction; **Chapter 5** - The branch of civil engineering which aims to minimize the effect which civil engineering activities have upon the environment is termed as sustainable civil engineering. The following chapter elucidates the varied aspects associated with this area of study such as green building, groundwater banking, seawall and rain garden, etc.

At the end, I would like to thank all the people associated with this book devoting their precious time and providing their valuable contributions to this book. I would also like to express my gratitude to my fellow colleagues who encouraged me throughout the process.

Sarah Woods

1

An Introduction to Civil Engineering

The branch of engineering which deals with the professional practice of construction, design and maintenance of infrastructure such as bridges, roads, canals, pipelines, dams, airports, etc. is known as civil engineering. This is an introductory chapter which will introduce briefly all the significant aspects of civil engineering.

Civil Engineering Systems

Civil engineering is arguably the oldest engineering discipline. It deals with the built environment and can be dated to the first time someone placed a roof over his or her head or laid a tree trunk across a river to make it easier to get across.

The built environment encompasses much of what defines modern civilization. Buildings and bridges are often the first constructions that come to mind, as they are the most conspicuous creations of structural engineering, one of civil engineering's major sub-disciplines. Roads, railroads, subway systems, and airports are designed by transportation engineers, another category of civil engineering. And then there are the less visible creations of civil engineers. Every time you open a water faucet, you expect water to come out, without thinking that civil engineers made it possible. New York City has one of the world's most impressive water supply systems, receiving billions of gallons of high-quality water from the Catskills over one hundred miles away. Similarly, not many people seem to worry about what happens to the water after it has served its purposes. The old civil engineering discipline of sanitary engineering has evolved into modern environmental engineering of such significance that most academic departments have changed their names to civil and environmental engineering.

These few examples illustrate that civil engineers do a lot more than design buildings and bridges. They can be found in the aerospace industry, designing jetliners and space stations; in the automotive industry, perfecting the load-carrying capacity of a chassis and improving the crashworthiness of bumpers and doors; and they can be found in the ship building industry, the power industry, and many other industries wherever constructed facilities are involved. And they plan and oversee the construction of these facilities as construction managers.

Coastal Engineering

Coastal engineering is a division of civil engineering responsible for the organization, conception, development and preservation of works on the shoreline. The objectives of these tasks involve management of shoreline erosion; improvement of navigation channels and harbors; protection against flooding brought on by storms, tides and even seismically triggered waves (tsunamis); improvement of coastal recreation; and management of pollution in nearby marine environments. Coastal engineering typically includes the development of structures, in addition to the transportation and probable stabilization of beach sand along with other coastal sediments.

Functions and Responsibilities of Coastal Engineering

- Erosion and Coastal Engineering: The majority of the world's population lives near coastal areas, yet these prized real estates are prone to the constant dangers of erosion. Erosion prevention and the protection of the coastlines are major responsibilities of coastal engineering. Coastal Engineers come up with structures of defense to protect against the devastation attributed to either erosion or storm waves.

  Some of the structures built by coastal engineers to protect against erosion are Breakwaters, Seawalls, Groynes and Revetments.

- Flood Control and Coastal Engineering: Coastal engineers develop defenses to preserve coastal residential areas from the consequences of storm flooding. With adequately engineered and developed defenses, the negative effects of the bad weather are reduced and therefore flooding is averted.

- Harbors and Coastal Engineering: Coastal engineers construct artificial harbors by installing breakwaters along with other marine structures, generating a safe refuge for vessels, wherever no natural harbor exists. These kind of artificial structures suck in the pressure of the waves, repelling it from hitting the vessels anchored in the harbor.

- Dredging Operations and Coastal Engineering: Coastal engineers are responsible for conducting dredging operations to retain a secure route for the transportation of vessels into or away from waterways and harbors. The functions and responsibilities of coastal engineering are crucial to the protection of our coastal environment and the operation of our commercial shipping industry.

Geotechnical Engineering

Geotechnical engineering is a practice that relates to the engineering behaviour of the earth and its materials. As a branch of civil engineering it is of great importance to construction activities taking place on the surface or within the ground, as well as to mining, coastal, drilling and other disciplines. Despite having considerable overlap, geotechnical engineering differs from engineering geology in that it is a speciality of engineering, whereas the latter is a speciality of geology.

Geotechnical engineers are responsible for evaluating subsurface and soil conditions and materials, using the principles of soil and rock mechanics. They are commonly appointed as consultants on construction projects. Engineers also examine environmental issues such as flood plains and water tables. By doing so, they are able to determine whether a particular site is suitable for a proposed project, and can inform the engineering design process with regard to how ground conditions can be made safe and effective for construction.

From geotechnical surveys, engineers will be able to evaluate the stability of the ground, including any slopes and soil deposits, assess any risks and/or contamination, and help to determine the types of foundations and earthworks that will be required. The potential for hazards such as landslides, earthquakes and other seismic activity can also be assessed.

Geotechnical engineers can be involved in 'ground improvement', in which soil is treated through a variety of different techniques to improve strength, stiffness, and/or permeability. Geotechnical engineering is also important in coastal and ocean engineering, in relation to building wharves, jetties, marinas and coastal defences, as well as foundation and anchor systems for offshore structures such as oil rig platforms. Engineers may also work on embankments, tunnels, channels, reservoirs, irrigation systems, and so on.

Construction Engineering

Construction engineering is a specialized branch of civil engineering concerned with the planning, execution, and control of construction operations for such projects as highways, buildings, dams, airports, and utility lines. Planning consists of scheduling the work to be done and selecting the most suitable construction methods and equipment for the project. Execution requires the timely mobilization of all drawings, layouts, and materials on the job to prevent delays. Control consists of analysing progress and cost to ensure that the project will be done on schedule and within the estimated cost.

Construction Equipment

Construction equipment is a wide variety of relatively heavy machines which perform specific construction (or demolition) functions under power. The power plant is commonly an integral part of an individual machine, although in some cases it is contained in a separate prime mover, for example, a towed wagon or roller. It is customary to classify construction machines in accordance with their functions such as hoisting, excavating, hauling, grading, paving, drilling, or pile driving. There have been few changes for many years in the basic types of machines available for specific jobs, and few in the basic configurations of those that have long been available. Design emphasis for new machines is on modifications that increase speed, efficiency, and accuracy (particularly through more sophisticated controls); that improve operator comfort and safety; and that protect the public through sound attenuation and emission control. The selection of a machine for a specific job is mainly a question of economics and depends primarily on the ability of the machine to complete the job efficiently and secondarily on its availability.

Hoisting equipment is used to raise or lower materials from one elevation to another or to move them from one point to another over an obstruction. The main types of hoisting equipment are derricks, cableways, cranes, elevators, and conveyors.

Excavating equipment is divided into two main classes: standard land excavators and marine dredges; each has many variations. The standard land excavator comprises machines that merely dig earth and rock and place it in separate hauling units, as well as those that pick up and transport the materials. Among the former are power shovels, draglines, backhoes, cranes with a variety of buckets, front-end loaders, excavating belt loaders, trenchers, and the continuous bucket excavator. The second group includes such machines as bulldozers, scrapers of various types, and sometimes the front-end loader.

Usually called a dredge, the marine excavator is an excavating machine mounted on a barge or boat. Two common types are similar to land excavators, the clamshell and the bucket excavator. The suction dredge is different; it comprises a movable suction pipe which can be lowered to the bottom, usually with a fast-moving cutter head at the bottom end.

Excavated materials are moved great distances by a wide variety of conveyances. The most common of these are the self-propelled rubber-tired rear-dump trucks, which are classed as over-the-road or off-the-road trucks. Wagons towed by a rubber-tired prime mover are also used for hauling dirt. These commonly have bottom dumps which permit spreading dirt as the vehicle moves. In special cases side-dump trucks are also used. Conveyors, while not commonly used on construction jobs for hauling earth and rock great distances, have been used to good advantage on large jobs where obstructions make impractical the passage of trucks.

Graders are high-bodied, wheeled vehicles that mount a levelling blade between the front and rear wheels. The principal use is for fine-grading relatively loose and level earth. Pavers place, smooth, and compact paving materials. Asphalt pavers embody tamping pads that consolidate the material; concrete pavers use vibrators for the same purpose. Drilling equipment is used to drill holes in rock for wells and for blasting, grouting, and exploring. Drills are classified according to the way in which they penetrate rock, namely, percussion, rotary percussion, and rotary. Specialized construction equipment includes augers, compactors; pile hammers, road planers, and bore tunnelling machines.

Structural Engineering

Structural engineering is, mostly, considered as a subset of civil engineering dealing with the design and analysis of buildings and large non-building structures to withstand both the gravity and wind loads as well as natural disasters. Besides, it may also cover design of machinery, medical equipment, vehicles or any other objects where structural functionality or safety are involved. Structural engineers must ensure their designs satisfy building codes.

Structural engineering came to existence when the humans first started to construct their own structures. It became a more defined profession with the emergence of the architecture profession during the industrial revolution in the late 19th Century.

In a practical sense, structural engineering is largely the application of Newtonian mechanics to the design of structural elements and systems that support buildings, bridges, walls (including retaining walls), dams, tunnels, etc.

Structural engineers ensure that their designs satisfy a given design intent predicated on safety (i.e. structures do not collapse without due warning) and on serviceability (i.e. floor vibration and building sway do not result in occupants criteria discomfort). In addition, structural engineers are responsible for making efficient use of funds and materials to achieve these goals. Typically, entry-level structural engineers may design simple beams, columns, and floors of a new building, including calculating the loads on each member and the load capacity of various building materials (steel, timber, masonry, concrete). An experienced engineer would tend to render more difficult structures, considering physics of moisture, heat and energy inside the building components.

Structural loads on structures are generally classified as: live loads such as the weight of occupants and furniture in a building, the forces of wind or weights of water, the forces due to seismic activity such as an earthquake, dead loads including the weight of the structure itself and all major architectural components and live roof loads such as material and manpower loading the structure during construction. Structural engineers mainly fight against the forces of nature like winds, earthquakes and tsunamis. In recent years, however, reinforcing structures against sabotage has taken on increased importance.

Structural Design

Structural design determines the type of structure that is suitable for a particular purpose, the materials to be used, the loads and other actions that the structure must sustain, and the arrangement, layout and dimensions of its various components. Structural design involves detailed calculations to ensure that:

- The structure is stable.
- All parts have adequate strength to resist the design loads.
- The structure as a whole will remain serviceable throughout its design life and able to perform its intended function.

Finally, structural design involves the careful preparation of drawings that will communicate the engineering design to the contractors who will build the structure.

Structural Analysis

Structural analysis is an integral part of structural design. It involves the calculation of the response of the structure to the design loads and imposed deformations that it will be required to resist during its lifetime. In structural engineering, 'deformation' refers to when an object is changed temporarily or permanently due to applied force. These calculations allow structural engineers to select the right materials for the structure, and to ensure that it will be suitable for the purpose for which it is being built.

Earthquake Engineering

Earthquake engineering is the science of the performance of buildings and structures when subjected to seismic loading. It also assists analysing the interaction between civil infrastructure and the ground, including the consequences of earthquakes on structures. One of the most important aims of earthquake engineering is the proper design and construction of buildings in accordance with building codes, so as to minimize damage due to earthquakes. It is the earthquake engineer who ensures proper design of buildings so they will resist damage due to earthquakes, but at the same time not be unnecessarily expensive.

Seismic Vibration Control Technologies

The purpose of these technologies is to minimize the seismic effects on buildings and other infrastructure by the use of seismic control devices. When seismic waves start penetrating the base of the buildings from the ground level, the flow density of their energy reduces due to reflections and other reasons. However, the remaining waves possess significant potential for damage when they reach the superstructure.

Vibration control devices assist in the reduction of the damaging effects, and enhance the seismic performance characteristics of the building. When the seismic waves penetrate a superstructure, these are dissipated by the use of dampers, or dispersed in a wide range of frequencies. Mass dampers are also employed to absorb the resonant wave frequencies of seismic waves, thus reducing the damaging effects. Seismic isolation techniques are sometimes used to partly suppress the flow of seismic energy into the superstructure by the insertion of pads into or beneath the load bearing elements in the base of the structure. Thus, the structure is protected from the damaging consequences of an earthquake by decoupling the structure from the shaking ground.

Research on Earthquake Engineering

In order to properly understand how buildings and structures can stand up to earthquakes, extensive research has also been conducted on earthquakes. In order to obtain an in depth knowledge concerning the initiation and behaviour of earthquakes, it is essential to ascertain the mechanical properties and frictional characteristics of the crust of the earth. Observations from space have clarified the complete cycle of earthquake, including the silent accumulation of strain, transfer of stress between faults, release of strain, and failure of faults. Measurements on boundary zones of tectonic plates have explained the interaction of faults across hundreds of kilometres. Study of the stress transients that take place after earthquakes will determine the possibility of future earthquakes at other sites in the system. These studies have provided scientific explanations related to earthquake engineering and resulted in revision of concepts and practical application.

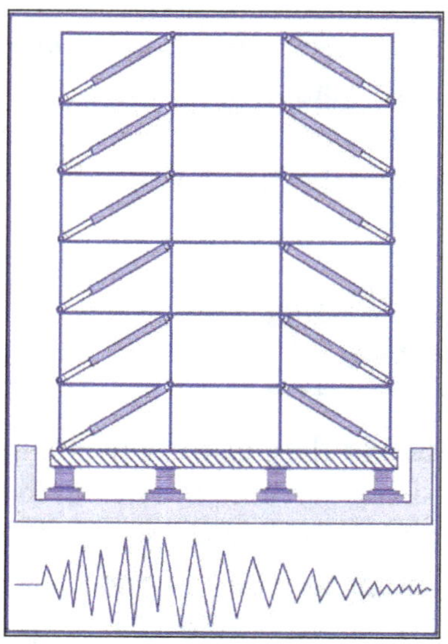

Transportation Engineering

Transport engineers plan, design and operate the large public and private infrastructure systems that connect our physical world. We need a broad range of continually evolving, large-scale transport infrastructure, including road, rail, air and water. Transport engineers quantify and optimise our mobility infrastructure networks to meet travel and freight demands, while ensuring safety, equity and sustainability, at minimal levels of congestion and cost.

Transport engineering has always been one of the essential civil engineering disciplines, impacting roadways, bridges, transit stations, airports and sea ports etc. Transport engineering has now developed into a multidisciplinary field spanning economics, politics, sociology and psychology, in addition to its core mathematical, engineering and computational principles.

Transport planning involves developing mathematical techniques for:

- Forecasting travel demand and planning to accommodate growth in demand.
- Determining improvements to the transport infrastructure.
- Reducing emissions.
- Reducing energy use.

Computational transport planning uses mathematical methods to predict, represent and quantify:

- The evolution of land use in cities.
- Travel attributes such as trip purpose.
- Travel decisions, including mode choice.

Planning models then examine the feasibility of projects and policies through cost-benefit and scenario analysis.

Transport Design

Transport engineers face multi-faceted design decisions when they are designing optimised transport infrastructure networks. These might relate to:

- The physical expansion of transport facilities, such as lane width or the number of lanes, for a roadway.
- The materials and thickness used in pavements.
- The geometry of a facility, such as a roadway, rail line or airport.
- Road pricing schemes.
- Deploying information-based technology.

In all design decisions, multiple performance measures, cost metrics and safety criteria must be considered and weighed.

Transport Operations

Transport operations, whether for road, rail, port or air traffic, are designed to minimise travel delays, improve safety, reduce emissions and enhance reliability, as well as taking other considerations into account. Transport operation decisions involve:

- Optimising traffic signals.
- Setting specific tolls.
- Designing traffic signs and markings.

With the development of new Intelligent Transportation Systems (ITS), transport engineers use tools including advanced traveller information systems (such as variable message signs), advanced traffic control systems (such as ramp meters) and vehicle-to-vehicle (V2V) communications to optimise the performance of the transport system.

Highway Engineering

Highway engineering is an engineering discipline which involves the design, construction and maintenance of Highway Roads & Systems, urban streets as well as parking facilities. Important aspects of highway engineering include overall planning of routes, financing, environmental impact evaluation, and value engineering to compare alternatives. Traffic engineering involves planning for the volumes of traffic to be handled, the methods to accommodate these flows, the lighting and signing of highways, and general layout. Older techniques include signs, signals, markings, and tolling. Newer technologies involve intelligent transportation systems, including advanced traveler information systems (such as variable message signs), advanced traffic control systems (such as ramp meters), and vehicle infrastructure integration. Similarly, pavement and roadway engineering involves setting of alignments, planning the cuts and fills to construct the roadway, designing the base course and pavement, and selecting the drainage system. Whereas, bridge engineering involves the design of highway bridges, retaining walls, tunnels, and other structures. Engineers in this specialization:

- Handle the planning, design, construction, and operation of highways, roads, and other vehicular facilities as well as their related pedestrian realms.
- Estimate the transportation needs of the public and then secure the funding for the project.
- Analyse locations of high traffic volumes and high collisions for safety and capacity.
- Use civil engineering principles to improve the transportation system.
- Utilizes the three design controls which are the drivers, the vehicles, and the roadways themselves.

Railway Engineering

It is a branch of civil engineering concerned with the design, construction, maintenance, and operation of railways. Railway engineering includes elements of civil, mechanical, industrial, and electrical engineering. Railway engineers handle the design, construction, and operation of railroads and mass transit systems that use a fixed guide way (such as light rail or even monorails).

Typical tasks would include determining horizontal and vertical alignment design, station location and design, construction cost estimating, and establishment of signalling & controlling system. Railroad engineers can also move into the specialized field of train dispatching which focuses on train movement control.

Railway engineers also work to build a cleaner and safer transportation network by reinvesting and revitalizing the rail system to meet future demands. In the United States, railway engineers work with elected officials in Washington, D.C. on rail transportation issues to make sure that the rail system meets the country's transportation needs.

Port and Harbor Engineering

Port and harbor engineers handle the design, construction, and operation of ports, harbors, canals, and other maritime facilities. A harbor (or haven) is a place for ships to enter and find shelter from storms or other natural phenomena. The modern harbor is a place where ships are built, launched, and repaired, as well as a terminal for incoming and outgoing ships. There are four principal classes of harbors; commercial, naval, fishery, and refuge for small craft. Harbor may be natural or artificial.

A port is a harbor with the necessary terminal facilities to expedite the moving of cargo and passengers at any stage of a journey. A good harbor must have a safe anchorage and a direct channel to open water, and must be deep enough for large ships. An efficient port must have enough room for docks, warehouses, and loading and unloading machinery. Geographically, a port or harbor is usually limited to a comparatively small area of usable berthing space rather than an extended coastline. Some ports along exposed coastal areas, for example, the western coast of South America, have little harbor area.

Airport Engineering

Airport Engineering encompasses the planning, design, and construction of terminals, runways, and navigation aids to provide for passenger and freight service. Airport engineers design and construct airports. They must account for the impacts and demands of aircraft in their design of airport facilities. These engineers must use the analysis of predominant wind direction to determine runway orientation, determine the size of runway border and safety areas, different wing tip to wing tip clearances for all gates and must designate the clear zones in the entire port.

Pipeline Engineering

Pipeline engineering embraces the design and construction of pipelines, pumping stations, and storage facilities. Pipelines are used to transport liquids such as water, gas, and petroleum products over great distances. Also, products such as pulverized coal and iron ore can be transported in water slurry.

Environmental Engineering

Environmental engineering is the development of processes and infrastructure for the supply of water, the disposal of waste, and the control of pollution of all kinds. These endeavours protect public health by preventing disease transmission, and they preserve the quality of the environment by averting the contamination and degradation of air, water, and land resources.

Environmental engineering is a field of broad scope that draws on such disciplines as chemistry, ecology, geology, hydraulics, hydrology, microbiology, economics, and mathematics. It was traditionally a specialized field within civil engineering and was called sanitary engineering until the mid-1960s, when the more accurate name environmental engineering was adopted.

Projects in environmental engineering involve the treatment and distribution of drinking water, the collection, treatment, and disposal of wastewater; the control of air pollution and noise pollution; municipal solid-waste management and hazardous-waste management; the clean-up of hazardous-waste sites; and the preparation of environmental assessments, audits, and impact studies. Mathematical modelling and computer analysis are widely used to evaluate and design the systems required for such tasks. Chemical and mechanical engineers may also be involved in the process. Environmental engineering functions include applied research and teaching; project planning and management; the design, construction, and operation of facilities; the sale and marketing of environmental-control equipment; and the enforcement of environmental standards and regulations.

Hydraulics Engineering

Hydraulics engineering is a field within the civil engineering discipline that addresses the control and management of water resources. Hydraulic engineering is the Science of water in motion and its interactions with the surrounding environment. Water plays a major role in human perception of the environment because it is an indispensable element. More importantly Human Life is totally dependent upon water. The technical challenges facing hydraulic engineers are formidable and sustained research efforts are essential.

The term 'Hydraulics' is related to the application of the Fluid Mechanics principles to water engineering structures, civil and environmental engineering facilities: e.g., canal, river, dam, reservoir, water treatment plant. Hydraulic engineering is the science of water in motion, and the interactions between the flowing fluid and the surrounding environment. Hydraulic engineers are concerned with application of the basic principles of fluid mechanics to open channel flows and real fluid flow hydrodynamics. Examples of open channels are natural streams and rivers. Man-made channels include irrigation and navigation canals, drainage ditches, sewer and culvert pipes running partially full, and spillways.

Hydraulic engineering links catchment hydrology and rainfall runoff to fluid mechanics of water runoff. In a catchment, a hydraulic structure must be analysed as part of the surrounding catchment and the hydrology plays an important role. Structural and hydraulic constraints interact, and the design of a hydraulic structure is a complex exercise altogether. First the system must be identified. What are the design objectives? What are the constraints? What is the range of options? What is the "best choice"? Its detailed analysis must be conducted and the engineers should ask: is this solution really satisfactory? During design stages, physical and computational models may be reliable 'tools' to compare the performances of various design options.

Urban Engineering

Urban Engineering is a civil engineering firm specializing in residential and commercial development planning & design, marine construction design and public works design. Urban engineering is the delivery of municipal engineering services, such as the supply of potable water, the removal and disposal of sewage, the protection of life and property from storm water, the disposal of solid waste, and the provision of road-based transport systems.

The delivery of each of these services requires engineering in respect of the long-term planning of the infrastructure or network, detailed design of the elements of the network, construction of each of these elements, on-going operation and maintenance, regular rehabilitation of the infrastructure and, when necessary, planning, design and expansion.

Equally important is the need to co-ordinate the design and delivery of all the services and to understand the social, anthropological, cultural, financial and political and other contexts in which delivery takes place.

References

- What-civil-engineering: civil.columbia.edu, Retrieved 27, Jan 2020
- What-is-coastal-engineering, civil-engineering-300995: ghhllc.com, Retrieved 23, July 2020
- Geotechnical-engineering: designingbuildings.co.uk, Retrieved 25, June 2020
- Earthquake-engineering-goals-technology-and-research-methods, structural-engineering-41861: brighthubengineering.com, Retrieved 08, August 2020
- Transportation-engineering: thecivilengg.com, Retrieved 08, Feb 2020
- Environmental-engineering, technology: britannica.com, Retrieved 05, April 2020

2

Geotechnical Investigations and Geological Considerations

The determination of the physical properties of soil below the Earth's surface in a proposed installation site by using different methods is known as geotechnical investigation. This chapter encompasses the geotechnical investigations for dam, reservoir and tunnel and geological considerations for construction, etc. All the diverse principles of geotechnical investigations and geological considerations have been carefully analyzed in this chapter.

Geotechnical Investigations for Dam

Dams have many effects on their local environments and have long been criticized for flooding landscapes, destroying or damaging downstream ecosystems, closing off fish migration routes, and in some cases forcing the evacuation of millions of people. However, for many years it was assumed that dams are a zero-pollution, climate-neutral way to produce electricity, since they burn no chemical fuel and emit no carbon dioxide (CO_2) in daily operation.

Classification of Dam

Dams are impervious to semi pervious barriers made across a river in a valley to impound water for slowing down river flow, for moderating flood, raising water level for canalization, hydro power generation etc. Dams are invariably associated with reservoir and depending upon the purpose of dam it may have many other features associated with it such as diversion barrage, canal headwork, guide bunds, access and service roads.

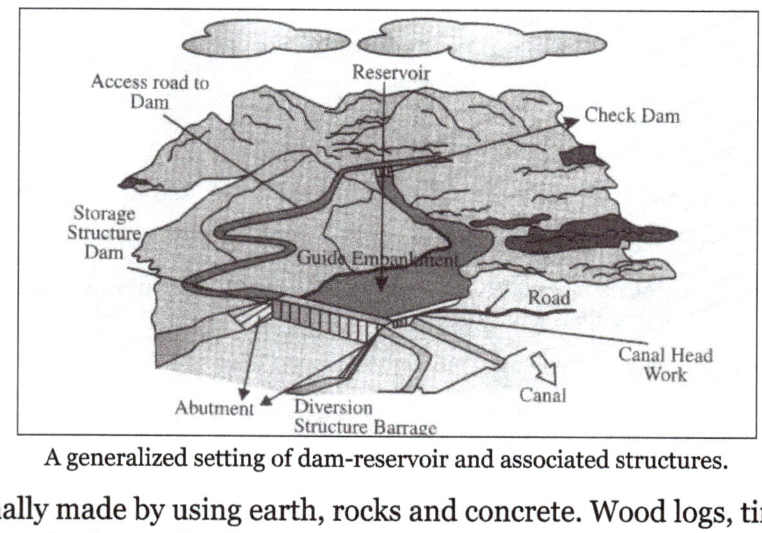

A generalized setting of dam-reservoir and associated structures.

Dams are traditionally made by using earth, rocks and concrete. Wood logs, timber sheets, plastic and steel planks can also be used. The bigness of a dam can be known by its length, and height.

The size of the associated reservoir in terms of its areal extent and the amount of water storage can also give an idea of the dam size. The dams are classified on the basis of their size, purpose and the material used in their construction.

Table: Classification of Dams

Size(meter)	<15	15-50	50-150	>150
Type	Small Dam	Large Dam	Major Dam	Big Dam
Purpose of Dam	To Stop Sediments and Mine Tailings	Diverting Water and Canalisation	To Impound Huge Amount of Water for Flood Control and Hydro Power Generation	
Type	Check Dams	Barrage/Weir	Multi-Purpose Dams	
Material	Earth	Rock	Concrete	Mixed
Type	Earthen Dam	Rock Fill Dam	Masonry Dam	Zoned Dam

The concrete or masonry dams are further classified as per their shape and size as Gravity Dam, Buttress Dam, Single and Double Arched Dam. Site, size and type of dam-reservoir, depend upon the presence of narrow river valley section, suitable foundation geology, availability of construction material and hydrological setup of the river system.

Gravity Dam

It is a huge structure made of either masonry or concrete with dam axis either straight or upstream curved or in combination of the two. Its cross section varies from triangular to trapezoidal shape with upstream side deck having gentler slope as compared to downstream one. The gravity dam owes its stability to its size, but exerts high magnitude stresses on rocks hence; its foundation should be laid on very strong rocks. The gravity dams require huge amount of construction material hence its availability need to be ensured. Gravity dam are generally provided with overflowing spillway in some portion of its length hence make it as two distinct unit as overflow and non-overflow sections. These separate sections are designed separately because of differing load conditions.

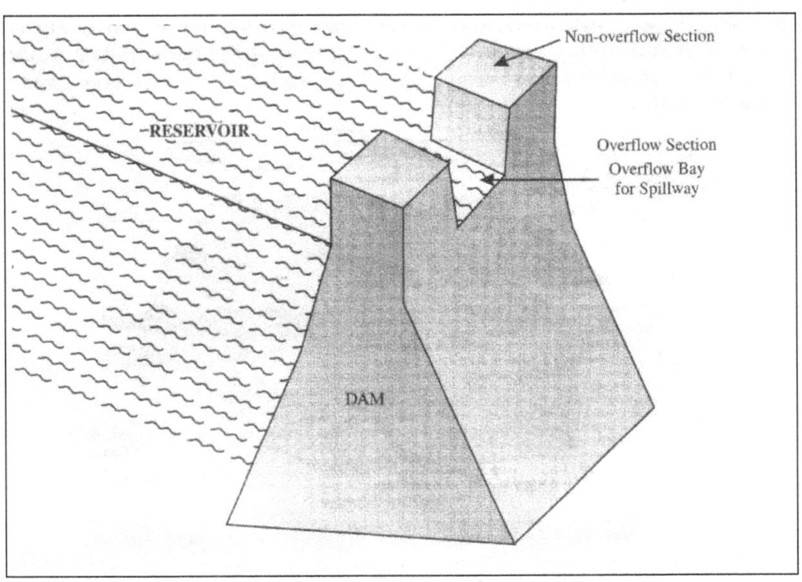

The typical trapezoidal cross section of a gravity dam.

Buttress Dam

In situations where strong and weak rocks are alternatively exposed the main dam is founded on strong rock, supported by perpendicular buttresses with a tapered joint called as corbel, in downstream direction which in turn are supported by struts to make it a complete monotonous unit so that the load is distribute on rocks uniformly. For example in combination of competent and incompetent rocks such as sandstone-shale, quartzite-slate, basalt-tuff whereon the main dam is made on competent rock supported by buttresses and struts spread over incompetent rocks. The dam can be made of rock masonry or concrete. The amount of construction material required is generally less than what is required in gravity dam; hence, buttress dam is economical.

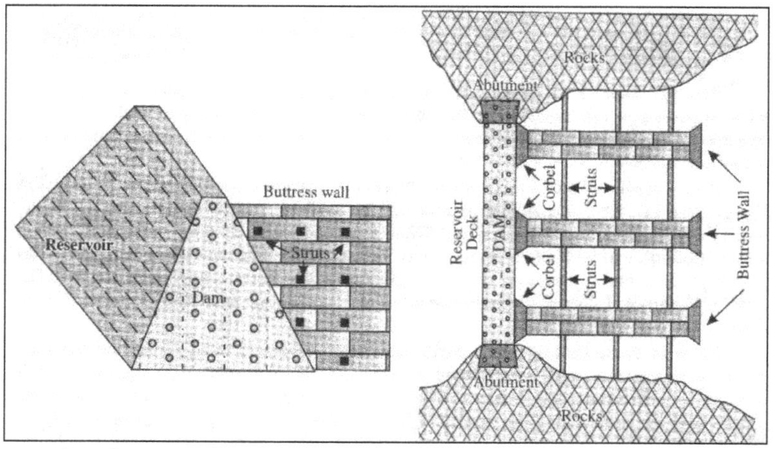

Left: Cross sectional view of a buttress dam showing main dam with deck and buttress wall.
Right: Plan of buttress dam, showing main dam with abutment, corbel, buttress walls and struts.

Arch Dam

Arch dams are thin as compared to gravity and buttress dams and require lesser amount of construction material. The arch dams are curved in plan towards upstream and may be of two types, Single or Double Arch Dam depending upon its curved nature only in plan or in plan as well as in elevation. The arch dams are made when thrust force of water is very high and it needs to be transferred towards abutments apart from foundation. Abutment rocks as well as foundation in this case need to be very strong.

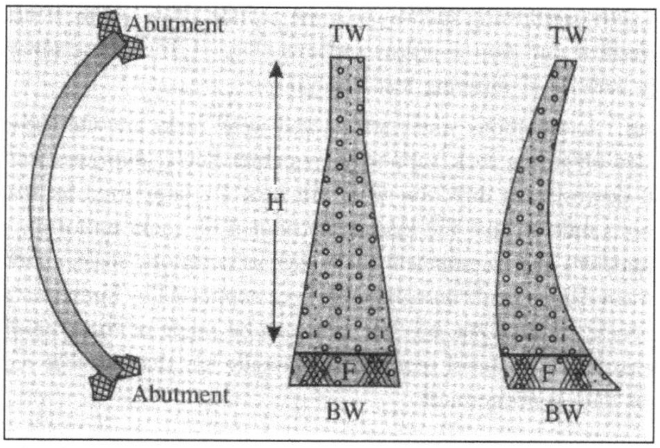

Arch dam as seen in (Left) plan, (Mid) cross section of single arch and (Right) double arch dam.
H – Height, TW - Top Width, BW - Base Width, F – Foundation.

Rock Fill Dam

Rock fill dams are made when procurement of concrete is expensive and sufficient amount of earth is also not available in nearby areas to make earth dam. Rock fill dams are best bet in areas prone to earthquakes and for regions prone to permafrost conditions. It can be constructed on moderately strong to weak rocks.

It may consist of only rocks with an impervious cushion on upstream side as shown in Figure. The rock rubble used should be available nearby as natural rock waste. Sometimes rock fill dams may have many different zones comprising rip rap as cushion, cushion to heel and toe made of rubble to prevent erosion, loose rocks forming the bulk of dam, hard rocks to protect the central impervious clay core, hearting or clay core to stop water seepage. The grade of slope of upstream side is low (1:1 to 1:1.5), as compared to downstream side (1:1.5 to 1:2.5).

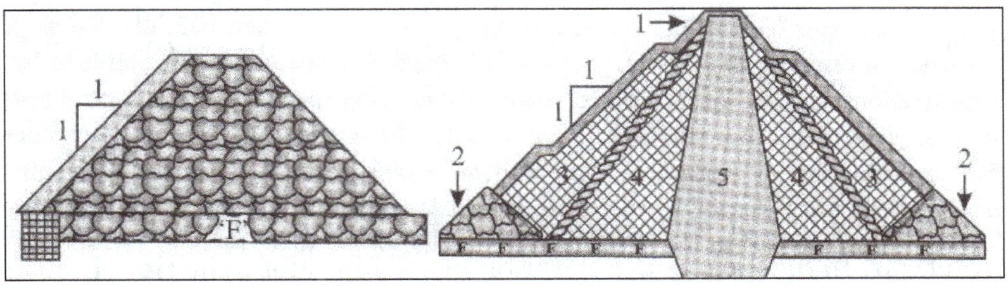

(Left) Simple rock fill dam with upstream cushion and cut off wall. (Right) Zoned rock fill dam with (1) riprap, (2) Heel and toe cushion, (3) Well graded rock rubble, (4) tough rock rubble and (5) impervious clay core or 'hearting'.

Earth or Embankment Dam

Earth dams are necessitated in case founding rocks are very weak or the river valley is very broad to make rock fill or masonry dam uneconomical. It is made by borrowed earth hence the availability of huge amount of earth is a prime requirement for earth dam. The earth should be strong enough to resist shear failure and impervious enough to prevent excessive loss of water through seepage. To make the dam stable the grade of slope should not be more than 2:1 in upstream side, it may be kept 1:1 to 1: 1.5 in downstream side. An impervious clay core in the middle of the dam is necessary to prevent water loss and rock or masonry lining in upstream face as a protection from erosion by reservoir water. Similarly proper compaction of earth on its optimum moisture content (OMC) is one of the prerequisite for a successful earth dam. Sufficient bond between foundation and abutment is of paramount importance to prevent piping and washouts. Post construction maintenance is a big issue in the case of earth dams because of settlement in crestal part, shear failure of upstream and downstream face, piping through the dam and development of cracks due to earthquakes.

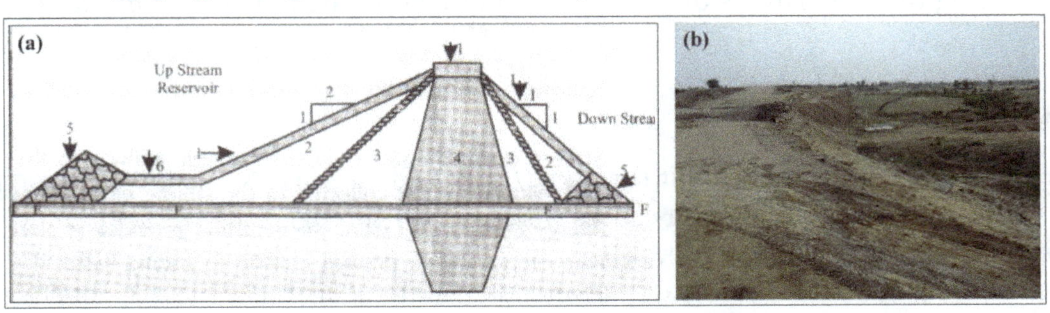

(a) Parts of an earth dam: 1- Riprap as cushion, 2- Bulk earth used for raising the dam, 3- Compacted earth, 4-Impervious clay core, 5- Rock cushion, 6- Berm. (b) An earth dam under construction.

Components of Dams

Main Dam

This is the main structure built across the river. The height of a dam depends upon desired storage capacity and the site conditions. The crest length of he dam depends upon topography at the dam site. The dam may be built of many different materials. The stored water is released from the dam as per requirements.

Flanks/Abutment

The rock mass on right & left banks of the river constitute abutments. Dam is joined with and supported by the abutments. In addition outlet tunnels, diversion tunnel, spillways are also placed in the flanks. The geology of the abutments has to be strong enough to enable placing various structural components without any risk. In addition abutments need to be of competent rock without any structural defects and lowest permeability.

Saddle Dam

The reservoir is usually formed by the main dam on one side and low/high hills on all other sides of the reservoir. In most cases the elevation of the hills along the rim of the dam is much higher than the reservoir maximum water level. In some other cases elevations of surrounding hills along a part of the rim/periphery of the reservoir is not high enough over a small section to completely contain the stored water and a saddle (low level place) is formed. Water can flow out through the saddle. A small embankment is then constructed at this low/saddle point to seal off the reservoir rim and is called as saddle dam. Example: Sukian dam and Jari dam for Mangla Dam project.

Diversion Channel/Tunnel

These channel or tunnel are constructed prior to dam construction such that river flow is passed around and away from the dam site through the diversion tunnels and that than dam site remain dry and accessible to construction at all time. The capacity of diversion structure is set such that most probable floods likely to occur during the construction period can be passed over without danger of overtopping of cofferdam and inundation of construction area. Necessary arrangements are made at d/s end for energy dissipation. These tunnels may be abandoned (plugged – Simly dam) after project completion or converted to irrigation/power/desilting tunnels. Diversion tunnel may not be provided (Mirani dam) and u/s coffer dam.

Cofferdam

These are small temporary dams built u/s and d/s of the dam site to make the construction area dry and workable. The u/s cofferdam causes water to flow through the diversion tunnel and the d/s cofferdam prevents backwater level to inundate the construction area. Coffer dam may be dovetailed in u/s part of dam (Mangla) or abandoned. Material used earth, rock, concrete etc. Arrangemnet are required for control of seepage across the coffer dam.

Spillway

This is a water release/conveyance structure to pass the large flood volumes safely across the dam without danger of overtopping of the dam crest. There would be one or more spillways usually at different levels (Service, additional, emergency). The lower spillway is used to release often occurring flood and regular inflows and is called as service spillway. It has usually more elaborate arrangements and may be free flowing or gated. The auxiliary or emergency spillway is set at or above normal conservation level and has fewer arrangements and is usually free flowing. This is used only during flood events of extra-ordinary nature. Fuse plug, rubber dam etc. may be used to delay water release and possible additional storage at the reservoir. The spillway may be a integral part of the main dam (mostly for concrete dams) or be a separate structure in the dam abutments.

Outlet Works

(a) Intake Structure/Tower: This is a structure to admit and control flow of water into the irrigation/power outlets. It would be a tower or inlet flush with reservoir side walls. Gates may be provided at u/s, intermediate or d/s end of the outlet tunnel. Necessary provision is made to keep the intake operation for long after sedimentation by having multiple water entry levels particularly for domestic supply purposes. Multi-level inlet openings may be used.

(b) Irrigation/Power Outlet Tunnel: This is a large water conveyance structure to release water to irrigation network and/or powerhouse turbines. The outlet is in the form of a tunnel dug or formed through the abutment/flank for earth/rock fill dams or through the dam body for a concrete dam. At the u/s end an intake is provided along with gates, trash rack. The tunnel design must eliminate risk of cavitation and/or aeration. Gates may be placed at u/s, d/s or intermediate location. The power tunnel is transitioned into surge chamber, penstock/scroll case etc. Energy dissipation structure may be provided at d/s end, if needed. Irrigation outlet may release into a canal or into the river if demand site is at distance from the dam. The intake level of the tunnel is kept below or at the dead storage level. Air vent is provided to minimize cavitation. Water cushon for vortex control are also provided.

(c) Low Level Outlet: A low outlet tunnel may be provided to flush sediments, draw water from below dead storage level under very drought condition, emptying of reservoir in emergencies, draw water during repair of outlet tunnel/gates, etc. The intake level is kept much lower than the intake level main irrigation tunnel. May discharge into stilling basin for spillways/outlet works or as a separate energy dissipation structure provided.

(d) Gates/Valves.

(e) Trash Rack, air duct for cavitation control.

(f) Drainage System: Dams are designed to store water with least seepage through the dam embankment and the foundation but seepage do occur. The drainage/seepage water also causes tremendous uplift pressure particularly at d/s half of the dam base. Features are included in the dam design to minimize seepage through the foundation and through the dam embankment and uplift pressure. Cut-off wall, sheet piles, slurry trench, etc.

- Grout Certain: An impermeable zone is created under the dam.

- Grout Blanket: Impermeable area is created u/s of dam.
- Pressure relief/Drainage Wells: Wells are installed at d/s area to pick and remove seepage water to reduce uplift pressure in the foundation area.
- Drainage gallery founation area: A horizontal/inclined gallery is formed in the body of the dam (especially in concrete dams) where water from drainage wells discharge into and is ultimately flow out of the dam body. It also intercepts leaks through dam body.
- Horizontal Blanket Drain: To intercept seepage lines at base of dam on d/s side.
- Chimney Drain: Vertical or inclined drainage filter layer (usually d/s of the impermeable clay core) to intercept seepage flow.
- Toe Drain: A drain is provided at toe of dam (homogeneous coarse fill) to intercept seepage flow inside the dam body.

(g) Preliminary Works: Trench provided at d/s of dam to intercept seepage flow lines. Gradient and seepage rate.

- D/S Trench: Civil works, infrastructures, buildings required to be provided before start of construction of main dam work. These include offices, staff housing, community buildings, water supply, approach road, client/consultant/contractor camp, labour camp, security arrangements, rest house, rail sidings, air strip, hele-pad, etc.
- Impermeable blanket to lengthen the seepage path and lower hydraulic.

(h) Hydropower Development:

- Powerhouse: Building to house turbine, generators, mechanical workshop, valves, draft tube, office, control room, visitor area, up transformer, etc for hydropower generation.
- Penstock: This is a large diameter pressure pipe used to deliver water to turbines.
- Surge chamber. To contain water hammer surge on plant load rejection/sudden shut-down.
- Switchyard: This is an area to install electrical equipment to change low to high tension power supply for further transmission.

(i) Slope protection/Riprap: Stone is placed on u/s & d/s dam slopes for protection against damage due to wave action, rain water, burrowing animals. Parapet wall may be used to protect dam top against sudden waves generated by strong winds, tsunami, etc.

(j) Dam Instrumentation: Various gages/instruments are installed in the dam body, foundations, spillway to monitor settlement, movement, stresses, pore water/uplift pressure, earthquake.

(k) Stilling Basin: To dissipate excess energy of diversion tunnel, low level outlet, irrigation tunnel, spillway, etc.

(l) Gallery/Shafts: These are provided in the dam body for access to interior of concrete dam body.

These are horizontal, vertical (with round stair ways), sloping.

(m) Operational Buildings: These are buildings required for operation of the dam and works. These include Office buildings, Rest House, Security buildings, Staff residences and other community buildings, gate control room.

(n) Temporary works: These are installations required for temporary use and are removed after project completion. These include contractor's camp, material processing, handling and stock area, machine room, casting yard, steel fabrication, labour camp, etc.

(o) Core: It is made up of impervious material and its object is to make the dam water tight. The particle size of material used in the core is small so that it can retain water and do not allow its further seepage.

(p) Shell: Shell is made up of pervious material (porous material). It gives strength and support to the core wall. The particle size of the materials used in shell is large i.e coarse material.

(q) Transition Filters: As the core contains fine material and the shell contains coarse material so there is a chance that both may get mixed into each other. These filters are installed in the dam body to stop the mixing of materials from either sides into each other. Transition filters are made of material which is semi-pervious in nature.

(r) Cut off trench: It is a trench dug to fill it with impervious material e.g. Rich Concrete.

(s) Sheet Pile Wall: Sheet pile wall is used in areas where it is required to stop water from flowing across. OR when the impervious layer is at very depth and if trench cannot be dugged out due to economic constraints then this sheet pile wall is used.

(t) Impervious Stream Blanket: Impervious stream blanket is laid in the foundation of the dam and its objective is to stop the seepage in the dam foundation.

(u) Rip Rap: This material basically belongs to the category of geotextiles and is used to stop upstream face from erosion. It is made up of large a stone which ranges from 1/2 m - 1 m.

(v) Horizontal Drainage: It is usually used if dam body is made of impervious material. The dam body may contain moisture already or due to very loss seepage, if this water remains in the dam body, pore water pressure is developed and effective pressure is reduced, so to remove that water this zone is used.

(w) Soil Turfing/Sod: To avoid erosion due to rain, snow fall, wind etc. soil turfing is used. It is also called sod.

(x) Crest/top: It is the most upper part of the dam which divides the upstream face and the downstream face. Its objective is to give access to vehicles over it.

(y) Free Board: To avoid overtopping of dam, free board is used.

Forces Acting on a Dam

The forces are considered to act per unit length of the dam. For perfect and most accurate design, the effect of all the forces should be investigated. Out of these forces, most common and important forces are water pressure and self-weight of the dam. In the design of a dam, the first step is the determination of various forces which acts on the structure and study their nature. Depending upon the situation, the dam is subjected to the following forces:

Water Pressure

Dams are mainly built to retain the water. So it has to be built to bear the water pressure loads. The well-known equation of the water pressure is as follows.

$$P = \gamma_w h.$$

Where P is the pressure (kN/m²), γ_w is the density of water (kN/m³), and h – is water height.

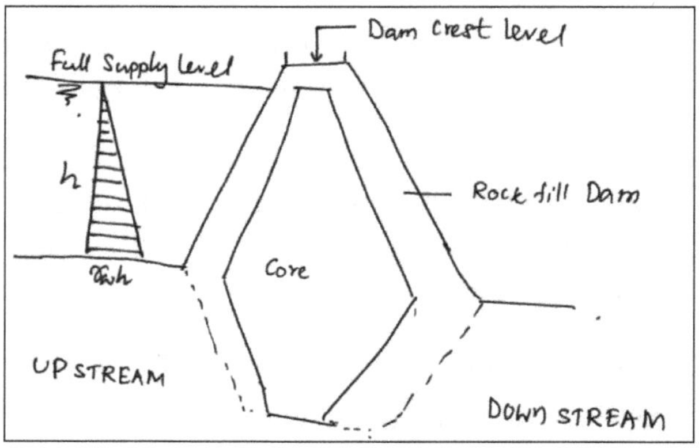

The above figure indicates the application of water pressure to a rock-filled dam. Similarly, the water pressure could be calculated for the other types of dams. We can calculate the force (F) applied on the dam as:

$$F = 0.5\gamma_w h^2$$

And forces will be active on one third height form which we considered the water height. The maximum water level shall be considered for the design. Generally, there are two types of water levels considered in dam design.

- Full Supply Level (FSL).
- Maximum Floor Level (MFL).

Full Supply Level (FSL)

It is the operation level of the dam. We maintain the water level generally at the full supply level. The height of the dame will be constructed above that level. Further, all the gate shall be operated at this level to meet the design requirement of the dam. For dams constructed only for power generation, its water level will be maintained at this level and when observed it increasing, more power generation will be done to maintain water level at FSL rather than sending it through the spillway.

Maximum Water Level (MWL)

The maximum water level is the water level that the dam will reach due to the increase in the water level beyond FSL in the case of high flood. The dam should be able to bear the rise of the water level up to the maximum water level and the spillway gates should have the capability to discharge the water without rising the water level beyond the maximum water level.

It shall plan to operate the gate between FSL and MWL and to maintain maximum water level as MWL during the operations. Further, more importantly, the dam should be designed for this water level too. When forces acting on a dam are considered, it shall make sure to consider water height up to the maximum water level to check the stability.

Uplift Pressure

Movement of the water under the dam body as seepage generates the uplift pressure on the dam. The stability of the dam is checked for overturning, sliding, and bearing. Uplift pressure on the dam is affecting the factor of safety against overturning and bearing. The effect on bearing could vary based on the type of ground that the foundation is constructed. Dams could overturn due to the lateral forces and uplift forces applied by the water. Therefore, uplift forces acting on a dam are considered very critical loads.

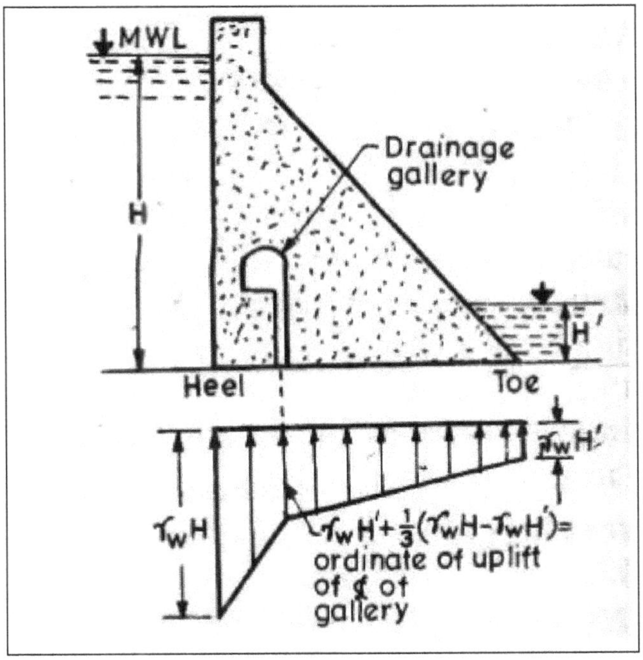

Uplift Pressure

The water level will vary under the foundation. Depending on the upstream water level and tail water level. The water pressure at the inspection gallery will vary due to the reduction in the water head. It could be due to the grout curtain or allowing water to drain through the drainage gallery. The above figure indicates allowing water to move through the drainage gallery to reduce the uplift pressure on the dam body.

In addition, grout curtains can also be constructed to avoid the movement of the water under the dam body. Further, curtain grouting is more popular in dam construction to minimize the seepages. The variation of the tailgate level shall be considered in the design carefully as it affects the uplift pressure considerably. The severe load case shall be considered in the design. When there is no drainage gallery or grout curtain variation of the head from upstream head to downstream head should be considered for the design.

Wave Pressure

Waves generated on the surface of the water due to the wind, apply additional loads on the dam. There are different methods to calculate the pressure on the dam:

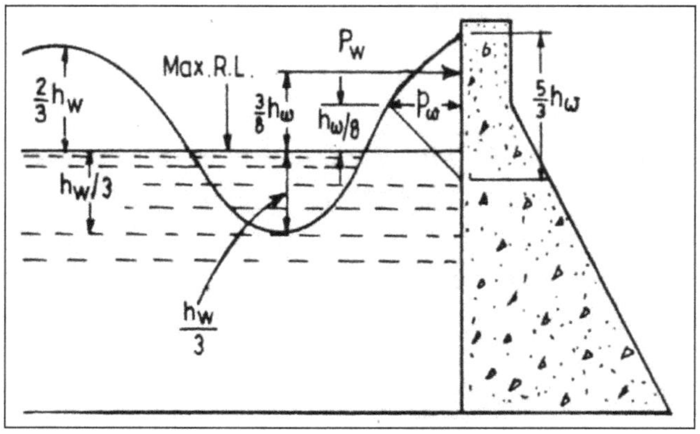

$h_w = 0.032\sqrt{VF} + 0.763 + 0.271(F)3/4$ for $F < 32$ km

$h_w = 0.032\sqrt{VF}$ for $F > 32$ km

Where, h_w – height of the wave in meters, V – wind velocity in km/h, F – straight length wave expands in km.

The maximum pressure P_w can be calculated as follows:

$P_w = 2.4\gamma_w h_w$ acts above $h_w/2$ above the still water level.

Further, the pressure distribution is considered as triangular and having the height 5hw/3.

Hence, total force(Fw) due to wave action can be calculated as:

$F_w = 0.5\,(2.4\gamma_w h_w).5h_w/3 = 19.62h_w^2$

Force will be acting 3hw/8 above the reservoir water level as indicated in the above figure.

Earthquake Forces

Depending on the location of the site and based on the return period, the magnitude for an earthquake is selected in dam design. A dam is such an important structure that costs a huge amount of money. Therefore, it should be safe enough to bear the loads applied to it. There should not any uncertainties in the design. Therefore, whether the dam is located in an earthquake-prone area or not, seismic forces acting on a dam are considered in the design.

Depending on the categorization of earthquake-prone areas, there are zones with specified magnitudes of earthquakes. It will base based on some probability or the return period. Specified magnitude shall be used if it matching with the return period considered in the design. If it is a high or less return period is considered, a suitable magnitude of earthquake shall be considered for the dam design.

There are two methods to calculate the dam or structures for earthquakes:

- Manual method or simplified method using initial force.
- More detailed analysis done manually or with computer-based software.

Manual Method or Simplified Method

Since we know the earthquake coefficient based on the data to be considered, we can calculate the acceleration. If we know the acceleration and mass of the structure, we can calculate the force easily.

Force = Mass x acceleration.

If we know the lateral force due to the earthquake we can calculate the overturning moment and factor safety. This is an approximation method with a lumped mass of the structure therefore a more precise method shall be used to get correct answers.

Water Pressure Due to Earthquake

Water pressure is varying when there is an earthquake. It is not static. The following figure indicates the application of water pressure during an earthquake:

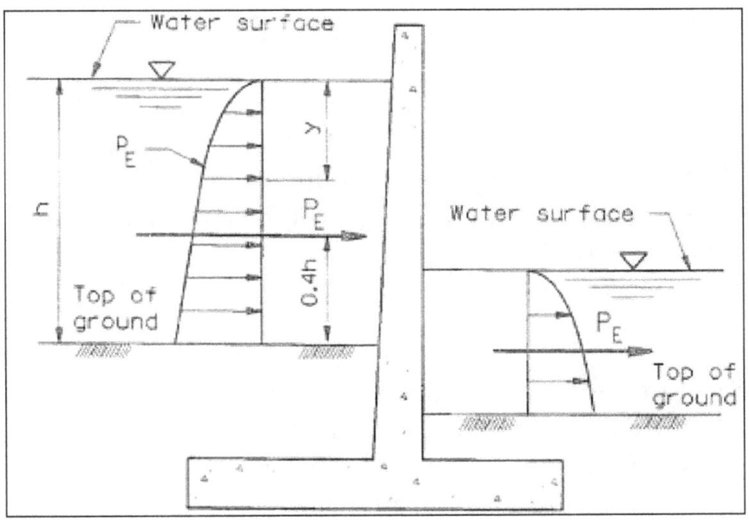

The above figure indicates the variation of water due to the earthquake. This pressure shall be added to the static pressure to find the final loading.

Wind Pressure

Generally, the wind pressure is not considered for the dams. When the dam is filled, a very high lateral load than the wind load will be applied to the dam. However, there is a portion of the dam exposed to wind beyond the full supply level. If required, we can consider that height for wind load application.

Silt Pressure

Silt deposits are inevitable in reservoirs. They reduce the depth of the dam and water storage. In addition, it becomes a burden to the dam too. With the time when there are floods, silt moves into the reservoir.

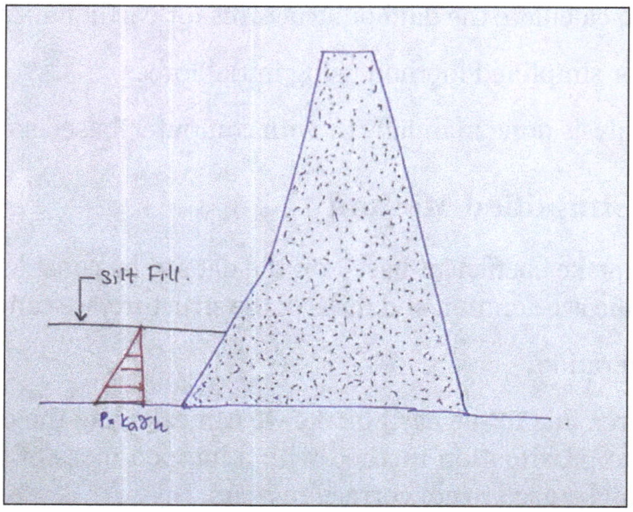

The submerged pressure applied by the silt can be calculated as shown above figure.

$$P = k_a\gamma'h.$$

Where, k_a – active earth pressure coefficient, γ' – submerged density of silt, and h – height of the silt deposit. Force can be calculated as follows:

$$F = 0.5k_a\gamma'h^2$$

Since the silt pressure is additional pressure applied to the dam, it shall be kept in mind to consider when design. Silt could make a considerable effect on the dam's overturning factor of safety if the dam venerable to have higher silt fill height. Therefore, a special assessment shall be made to know the forces acting on a dam due to silts.

Thermal Loads

Thermal loads are very critical for concrete dams. During the construction and after the construction effect to the temperature variations shall be monitored. If it exceeds the limiting values, there will be issues in durability of concrete. The rise of the concrete due to the heat of hydration shall be limited in concrete structures to avoid cracking. Further, attention shall be drawn on thermal cracks due to the variation of concrete temperature.

The following combination of loading shall be considered when the design for the crack width of the concrete:

- M + T+ T2
- T1 only
- T1 + T2

Where,

M – Bending Moment

T – Tension Force

T1 – Rise of temperature from ambient due to heat of hydration

T2 – Seasonal variation of temperature from ambient temperature

Impounding Stresses

The impounding of a dam is filling it for the first time. Impounding is very critical for rock fill dams, earth dams, or similar types of dams. When the water level rises, there will be stresses in the core with the saturation. Therefore, there are limitations in the rate of filling. It could depend on the specification of the project. It shall be allowed to gradually raise the water level in the reservoir.

Ice Pressure

In countries where there is variation in temperature due to seasonal effects. Water becomes ice. When they melt and expands a pressure will b applied to the dam. It could be in the range of 250 – 1500 kN/m2.

Self-Weight

The self-weight of the structure need to be taken into account when it designs of stability in serviceability limit state and ultimate limit state design. It is very easy to calculate the self-weight of a concrete dam. It is the density of concrete multiply by volume of concrete. However, when we need to calculate the weight of the rock fill dam, more effort is required as there are different types of layers with different densities.

Causes of Dam Failures

As mentioned in above section, any force, which undermines the downward acting vertical forces, will lead to the failure of the dam. This can be due to excessive pore water pressure in dam foundation, high buoyancy due to submergence of dam, high lateral pressure due to sudden filling of reservoir or high wave action in response to strong wind and storm.

The masonry dams fail mainly due to sliding in the foundation of embankment due to presence of weathered and weak rock material, improper keying either of the foundation or due to presence of clay, salts and other weak material in joints. Other common cause is sudden lowering of vertical downward acting forces by overtopping or by high lateral pressure due to excessive water in reservoir caused by inadequate spillway is another cause. Development of cracks due to differential settlement will also make the dam weak leading to its failure. Most of the masonry and concrete dams fail within few days to few months of their completion but plain or reinforced concrete dams (PCC/RCC) may develop fractures tens of years later due to alkali aggregate reaction, hence need proper surveillance and corrective measures.

Rock fill dam fail mainly due to caving caused by settlement below the foundation in case rocks are not strong. Sliding within the body of the dam is another important way of failure due to material inhomogeneity in terms of size, shape and composition of rock blocks. Sudden increase in lateral pressure or development of permafrost condition may cause washouts. The failure of earth dam may have many different reasons such as seepage and piping through the dam, embankment or foundation-causing washout.

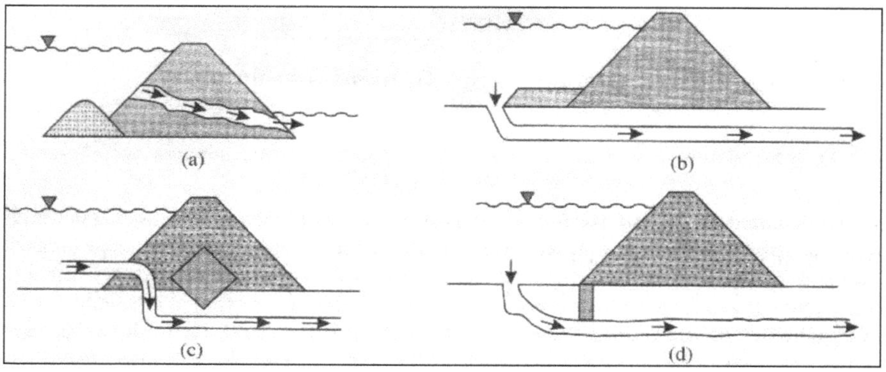

Piping through various part of the earth dam (a) through the main dam, (b) below the berm, (c) below clay core and (d) below cut-off wall.

Seepage of water through slow permeation can make its front as the failure surface similarly permafrost front. Differential compaction of the earth, sudden drawdown and seismic activity may also cause earth dam failure.

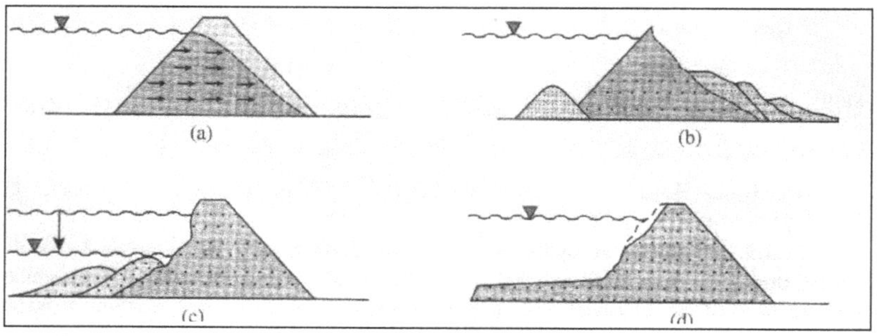

(a) The seepage of water through the earth dam or freezing of moisture in permafrost condition, separates it into wet and dry parts. (b) Downstream shear failure along the seepage or permafrost front. (c) Shear failure due to sudden drawdown of reservoir water in upstream side. (d) Shear failure due to earthquake and liquefaction.

A dam can fail due to many different reasons and shortcomings in geological and geotechnical investigations on one hand and poor design on the other. Presence of Weak rock mass or unfavourable geological conditions if not properly identified and corrected then will seriously affect the stability of the dam. Engineering geologists should explore extensively and intensively to ensure proper geological investigations. Any shortcoming in geotechnical investigations for foundation and abutment will make a dam inherently weak. A geotechnical engineer should not leave any stone unturned to make the very base of the dam strong. Improper design of the dam body will mar the very purpose of the dam. The structural engineers should design the dam with utmost care using state of the art design methods taking in view all the negative and positive points identified during geological and geotechnical investigations.

Table: Statistics of major causes of dam failures all over the world.

Causes	%	Causes	%
Foundation failure	35	Act of War	3
Inadequate Spillway	25	Embankment Slips	2
Poor Construction	15	Defective Materials	2
High-Pore Water pressure	10	Incorrect Operations	2
High Rate of siltation	5	Earthquakes	1

Suitability Factors

Suitability of Hydro-Meteorological Factors

If it is not solely for the flood control, then water is the most essential commodity for all purposes of dams. The availability of water to be used for irrigation, hydropower generation warrants ample amount of water in reservoir and to ensure this adequate study is a must to predict stream flow throughout the designed life of the dam.

- Climatic Setup and Weather: The overall availability of water to a river on a large time scale will depend upon the climatic belt, it is flowing into. The weather of that area will control the seasonal and day-to-day variability in the amount of water flowing into a river. Hence, the long and short term database of average annual rainfall/snowfall for last 100 years or more, maximum rainfall and minimum drought stage, periodicity of storm recurrence at 10, 25, 50 and 100 year level, average annual time span of dry season etc. are important meteorological parameters should be known and analysed prior to selection of river to be tapped or in deciding the size and type of the dam.
- Hydrological Setup of River: The database of stream to be damned in terms of average annual discharge, maximum flood level and minimum water flow are very important which apart from meteorology of the area will also depend upon the drainage density, basin shape and stream order. The grade of slope of mainstream as well as of its contributories is also important, as this will decide the rate by which water will come into reservoir. In an event of erratic rainfall the rate at which reservoir is going to fill up has to be taken into account in designing the spillway and the freeboard, to avoid overtopping.
- Carrying Capacity: All the rivers carry sediment with them. The total amount of sediment a river is carrying with it at a place is termed as carrying capacity of that river. The river distributes the sediment all along its course. When the river is damned, this sediment will now be deposited in the reservoir, termed as siltation. The siltation will take place in space provided as dead storage and as and when the total capacity of dead storage is taken over by the sediment then any extra siltation will eat out the space meant for reservoir water, which is unwarranted.

Suitability of Geomorphological Factors

Geomorphic setup of a river basin develops in millions of years and attains its dynamic equilibrium. A river gets water in its upper most reaches from snow/ice melt, rainwater through myriads of low order streams, termed as source of a river in its proximal part termed as catchment area. River by lateral, vertical and head ward erosion carves out valleys of different shapes and sizes and drain an area along with its tributaries termed as drainage basin. It empties itself in a lake or sea in its distal part at its mouth depending upon its being a river with internal or external drainage respectively. The water and sediment collected in catchment area and on its way to mouth are distributed throughout the drainage basin.

Depending upon the site of proposed dam the river will be severed from its distal parts, only part of water and only suspended sediment will now go through the spillways while rest of the water and maximum amount of sediment will be held back in the reservoir itself. Now for the river upstream of the dam though original source region may remain intact but it will have new mouth i.e. reservoir while for river downstream of the dam will have the new source i.e. spillway, while it may retain the original mouth. This change in the dynamic equilibrium of the river may have far reaching consequences and may lead to excessive deposition in upstream and erosion in downstream. Hence, detailed geomorphological investigation of catchment, reservoir and dam site area is a must.

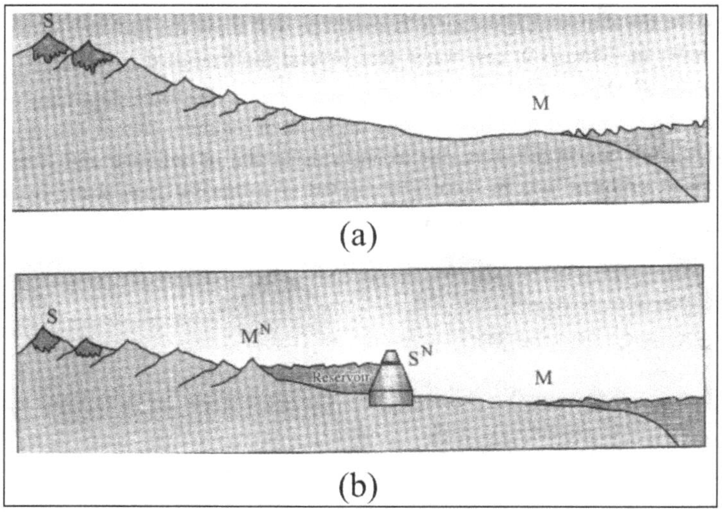

(a) The graded profile of a river showing its source (S) in catchment area and mouth (M) near the sea before the construction of the dam-reservoir. (b) New source and mouth (S^N/M^N) of the same river after construction of the dam reservoir.

Geomorphology of Catchment Area

Geomorphology of catchment in terms of slope characteristics apart from shape and size of the catchment area are important in deciding the total amount of water available to a river. The other geomorphic factors such as drainage pattern, drainage density, stream order and sinuosity of river should also be properly worked out as they are going to control the velocity with which available water will come into the reservoir.

Geomorphology of Reservoir Area

The reservoir behind the dam once filled may have different size and shape depending upon the geomorphology of the area as well as the length and height of the dam. The reservoir rim formed by the topography of the area will result into its linear, oval, semi-circular to irregular shapes and will have different pattern of forcing the dam. The danger of landslides in reservoir increases with increasing slope. The scars of previous landslides and slopes showing creeping should be identified and mended. The reservoir water may also lead to increase in groundwater level leading to lowering of frictional resistance along the discontinuity surfaces aggravating landslides.

Geomorphology of the Dam Site

The most important geomorphic consideration at dam site is the narrow river valley section. The narrow is the river valley smaller will be the length of the dam and lower will be the cost. Sometimes this becomes the only criteria for the selection of dam site to economize its construction, leaving (ignoring) other factors as the case is with Tehri Dam (Uttarakhand) wherein the city of Tehri was sacrificed for a site which gave length of the dam only 656m, almost half of its width which is 963 m at its base, a rare case. The height of the shoulder hills is also important, on to which the two sides of dam are going to rest i.e. the abutment foundation. It should be well above the dam height so as to prevent spilling of water in sideways in case water goes beyond the freeboard.

Thickness of Alluvial Fill

The river valley may have varying width depending upon its stage and antiquity. Accordingly, it may have varying thickness of sediment deposit, termed as valley fill. As rivers shift their course sideways, due lateral accretion of sediment and vertically due to deposition of sediment layer by layer. The coarsest of sediments (gravels and coarse sand) are deposited in its channel and are left back as shoe string deposits. When dam is made and reservoir filling starts, water percolating downward through sediment deposits find easy to escape through these so called highly porous and permeable zone of sediments termed as buried channels, resulting not only loss of stored water but also problem of seepage in foundation leading to solution and development of pore water pressure. Hence such buried channels are to be explored and identified (geophysical exploration), sealed and be abutted and in no case they should lie below the dam foundation.

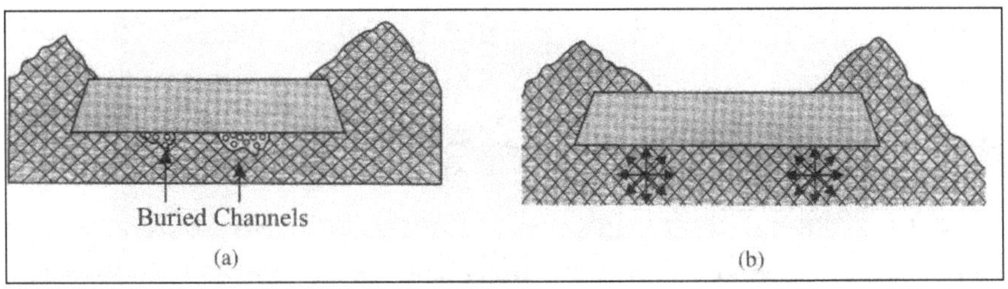

(a) Buried channel deposits acting as conduit for water seepage "Buried Channel" below the dam;
(b) Pressure created by water in fractures and rock pores causing "Pore/Fracture Water Pressure" acting against the vertical loads exerted by dam and reservoir water.

Suitabilty of Geological Factors

As we know that formal application of geological knowledge in civil engineering started after the analysis of many dam failures, took place in USA and Europe. A detailed investigation regarding the geology of the reservoir and dam site have since become mandatory before embarking upon the construction. The important issues to be taken into consideration are related to rock mass properties, their attitude and presence of major deformation structures. As the reservoir and dam setup covers a considerably big area it is likely to bring in as many geological issues. It is also a well-known fact that each site may have its own geological peculiarity hence the elements of uncertainties will always be there. The dam site is especially important and it is only after the excavation for foundation and abutment is completed real picture of geology will emerge to be considered in final design.

Rock Type

Depending upon the size of a dam its safety increases many fold if there is only one rock type present in its foundation and abutments. Igneous rock bodies being very large three dimensionally, offer such condition more than often.

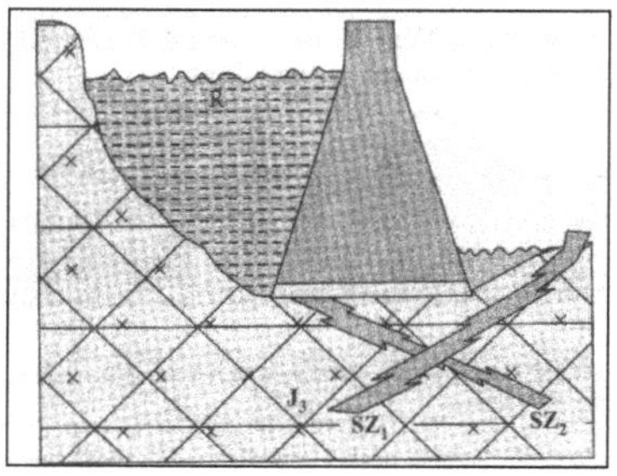

Granite, a plutonic igneous body covering entire dam and reservoir area.

Metamorphic and sedimentary rock bodies may or may not be large enough to host all elements of dam within a single rock body. The overall strength of rock mass will depend upon its weathering state and other rock mass properties but it is the bearing capacity, which will be an important factor in deciding the type of dam. The gravity dams, being huge in size are only made on very strong rocks. Hence, igneous and non-foliated metamorphic rocks will be most ideal. Volcanic igneous rocks if thick and without intervening ash beds and sandstones with good strength can also host foundation of gravity dams safely. Rock fill dams are ideal for earthquake prone regions and also are the best bet in areas with rocks of moderate strength. Foliated metamorphic rock gneisses, thickly bedded sandstones and strong limestone in comparatively dry regions have been found to perform well. All other rocks, which are thinly layered and weak such as volcanic flows with ash beds, tuff, schist, phyllite, slate, marl, thinly bedded sandstone, limestone etc., are suitable only for earth dams.

Weathering State

The weathering state of rocks will depend upon its composition, intensity of jointing, climatic set-up of the area and geological age. In no case weathered rock be made as foundation ground for any type of dam. In fact for large dams only fresh rock surface should be taken into consideration and if not naturally available then should be excavated up by scrapping.

Attitude of Rocks

In the case of stratified rocks the orientation and thickness of rocks in relation to the dam axis and resultant force direction plays a very important role. The horizontal rocks or rocks dipping by less than 50 do not offer good foundation condition as it may come in parallelism with resultant force direction. If rock beds are thicker than the dam height the whole dam with foundation will remain in one rock. However, if the thickness is less, then rocks in foundation and abutment may differ. In such cases, care should be taken in deciding the foundation of the dam, which should be placed in the strongest rock bed.

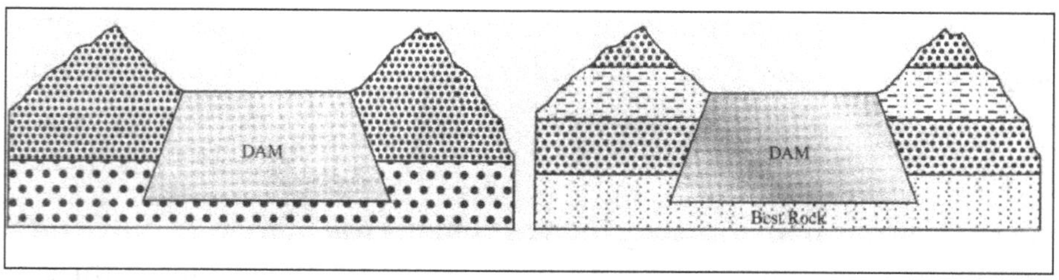

(Left) Whole dam unit lying in one horizontally laid rock body. (Right) Placement of dam in the strongest rock body in case of multiple horizontal rocks.

The vertical rocks may get disposed in three different fashions with respect to dam axis. If striking parallel to dam axis and thickness of rocks are more than the width of dam axis then only one and if less then more than one rock will be encountered in foundation and abutments. The resultant force direction will make high angle to vertically dipping rocks, which is a favourable condition. In case rocks are striking perpendicular to dam axis then there are chances of having more than one rock type will become more due to the fact that rocks may not be as thick as the length of the dam. This situation is also not favourable as resultant force direction will lie in plain of rock contacts. If rocks are striking obliquely then the rocks may be disposed in between these two extreme cases and will give fair foundation conditions.

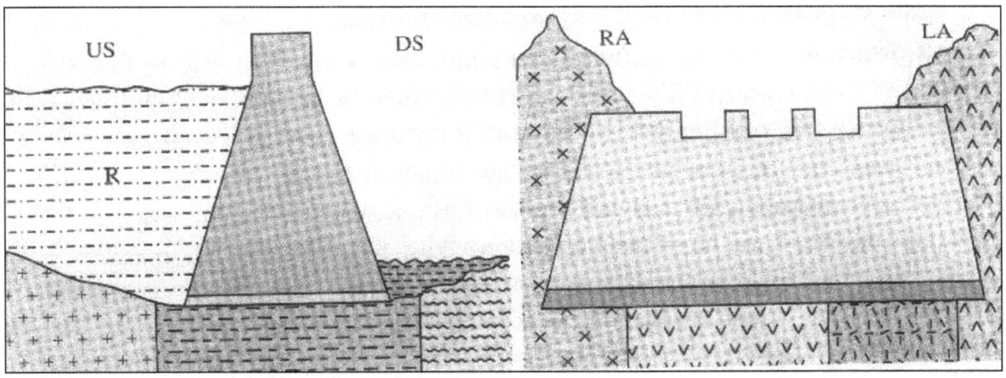

(Left) Entire base width of a dam showed resting on a thick vertical rock striking parallel to the dam axis. (Right) Multiple vertical rocks lying below the dam in case rocks are striking perpendicular to the dam axis.

The inclined rocks will show varied nature of disposition depending upon its strike and dip with respect to dam axis. The rocks if strike parallel to dam axis then it may dip towards up or down stream directions. The rocks dipping upstream will be the most favourable situation as resultant force direction will be perpendicular to the dip directions. Rocks dipping downstream will result into most unfavourable situation. This may become worst if the rocks are dipping less than 450 as resultant force direction will come in parallelism with rocks dip. The case in which rocks are striking perpendicular to the dam axis and rocks dipping either towards right or left abutment of the dam and rocks dipping obliquely to the dam axis with different dip directions may pose fair condition.

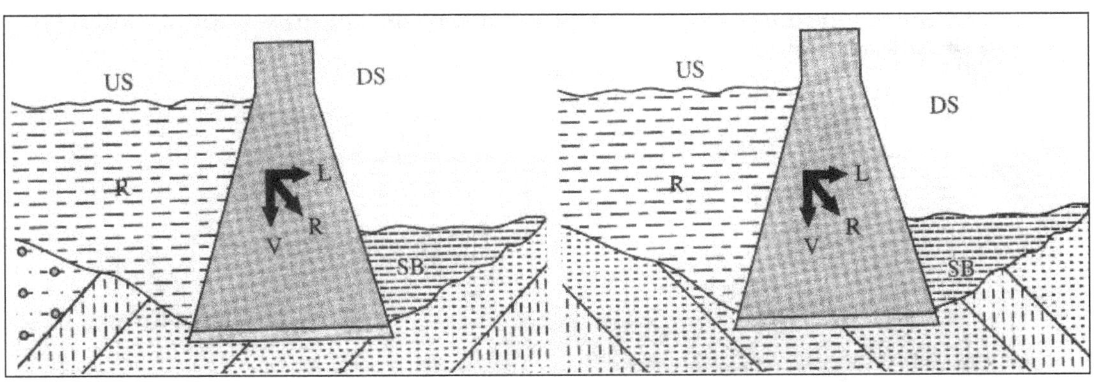

Figure Shows (Left) Inclined rocks striking parallel to the dam axis and dipping in upstream direction, see only one rock in foundation due to its thickness more than the foundation base. (Right) Inclined rocks striking parallel to the dam axis and dipping in downstream direction, see more than one rock in foundation due to the thickness of rocks less than the foundation base.

Rock Structure

The rock sequence over which dam is proposed may be folded, faulted and sheared to various degrees. The presence of folds or faults or both, will not only result in repetition and omission of rocks but will bring in many other issues and problems.

The construction of dams will be effected by attitude of folding with respect to tunnel length and width. Depending upon the thickness of rocks and wavelength of folding repetition of rocks will take place and rocks will be oriented differently across and along the dam length. In the case of fold axis running parallel to the dam axis, the dam foundation may run over an anticline or syncline. The position of dam foundation can accordingly be selected to get maximum benefit of rock inclination. Another aspect is of groundwater seepage along the folded layers; the water will move away from the dam in case of anticlines and will move into in the case of synclines. It is suggested that if the rocks involved in folding are thick enough then the best option is to have dam between the anticline and syncline axis i.e. in the limb dipping in upstream directions. If the fold axis is perpendicular or oblique to the dam axis then depending upon its wavelength one or more than one, anticline and syncline may come across with varying conditions of rock orientation.

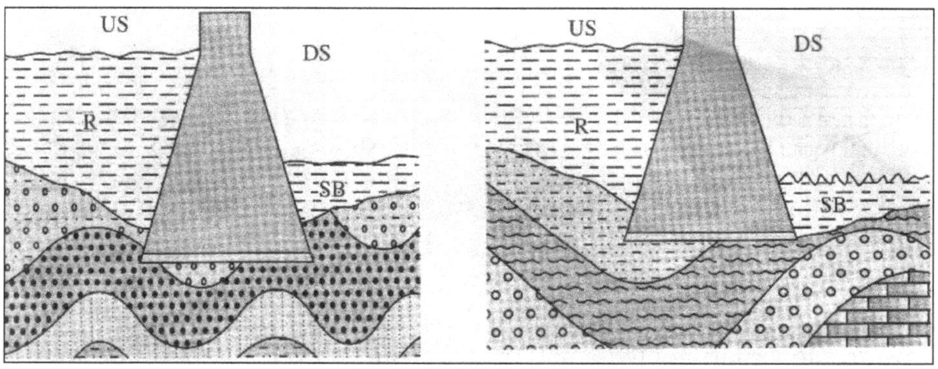

The folding of rocks with fold axis parallel to the dam axis and wave length less (Left) and more (Right) than the base width of dam and accordingly placement of dam to get upstream dip.

Fault and fault zones should be avoided if it is live. Even if it is a dead fault then too it should be identified in terms of its type and disposition with reference to dam axis and dam should cross a fault in perpendicular fashion to have its run at minimum in dam foundation. Faults lying in regions of Seismic Zone IV to V, such as Himalayan Mountain System should be seen with caution and placing of dam may be decided accordingly.

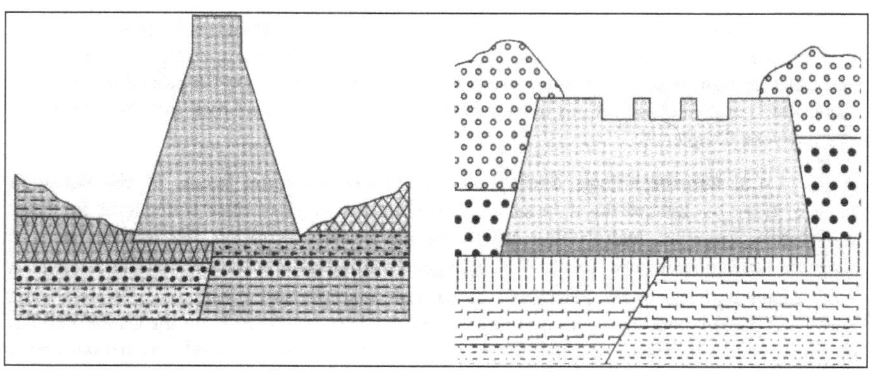

(Left) A fault oriented parallel to dam axis will run all along its length, a dangerous situation. (Right) A fault-oriented perpendicular to dam axis will run only along the width of dam, somewhat manageable situation.

The presence of crushed material along the fault or fault zone (gouge/mylonites/breccia), being weak should be excavated out and be backfilled by concrete. The crushed zone may also act as medium for groundwater storage and movement. Its extension in reservoir area or exposure to ground surface may lead to percolation of water in to foundation of dam hence it should be identified and sealed. There are other issues related to fault such as repetition and omission of rocks resulting into change of lithology across the dam.

The presence of unconformity will bring in two different sequences of rocks across it. The presence of residual conglomerates and/or palaeosols etc. will make it undesirable at least near or below the foundation dam.

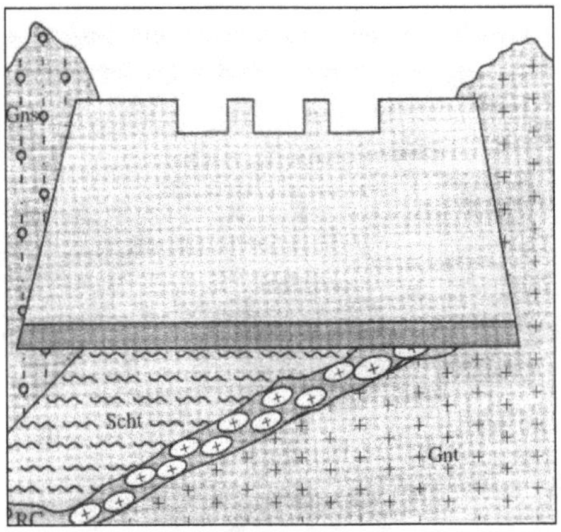

Dam across an unconformity bringing granite (Gnt) and metamorphic rocks comprising granulite (Gnlt) and gneiss (Gns) together. Also, see the residual conglomerate (RC).

The shear zones and Joints being the most important in controlling the overall strength of rock mass, on to which dam is to be constructed, are destined to play the most important role. Shear zones can act as weak zones in rocks along which rocks are sheared and pulverized and need to be excavated out and backfilled by concrete.

As joints are ubiquitous they are going to play the most important role in terms of numbers of joint per unit area, spacing between the joints, opening along joint aperture, joint condition and water in joint planes. Based on strength of rocks and various properties related to joints the RMR and Q system classifications of rock mass can be used to predict rock mass behaviour and strength.

Geotechnical Investigations for Reservoir

Places where fluids collect are called reservoirs. The most common fluids found in reservoirs are water, hydrocarbons, and gas. Reservoirs can be natural or artificial (human-made). Examples of natural reservoirs include hydrocarbon reservoirs in rock formations and water reservoirs that occur behind naturally occurring dams. Artificial dams mostly involve water, such as the reservoirs behind hydroelectric dams.

Within an energy context there are different types of reservoirs and the word reservoir can refer to:

- Reservoir - Oil and Gas: a formation of rock in which oil and natural gas has accumulated.

- Reservoir - Water: a human made structure that stores water in a defined volumetric area, most specifically a hydroelectric reservoir.
- Natural gas storage: extracted natural gas is sometimes injected into depleted reservoirs, either for storage or to increase the pressure in the reservoir to make extraction of other substances easier.

Reservoir Examples

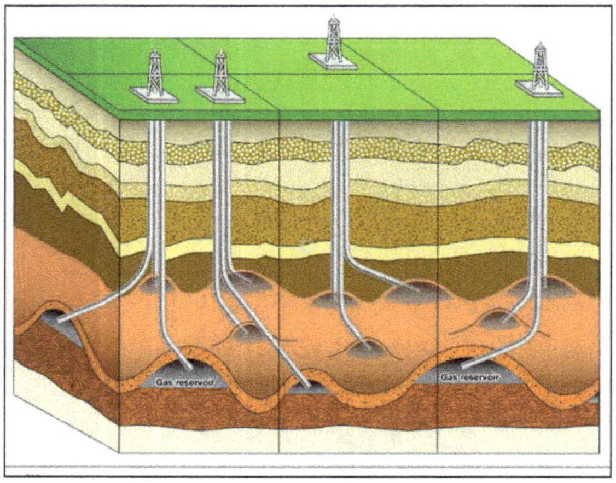

Gas reservoirs are located deep underground.

The reservoir behind the Hoover Dam is called Lake Mead.

Geomorphology and Geology of the Reservoir Area Reservoir behind the dam serves as storage for water to raise water level for flood moderation, irrigation, hydropower generation and also for recreation. However, impoundment of water in such an enormous quantity also has geo-environmental consequences.

Shape and Size of the Reservoir

As the main objective of the reservoir is to store water, which will depend upon the space available behind the dam depending upon the geomorphology of the reservoir area, which in turn will control its storage capacity and yield? The reservoir body may take different shapes such as irregular, oval, elliptical to linear and accordingly will have varying depths. The maximum possible storage will depend upon the height of the dam while yield will be controlled by annual average inflow of water and losses due to evaporation and seepage. The amount of water supplied to the reservoir will be controlled by the size and shape of the catchment area, drainage density and slope of the river and its tributaries. The rate of water ingress by the river in to reservoir and designed discharge through spillway is very important to keep maximum reservoir level as per the design.

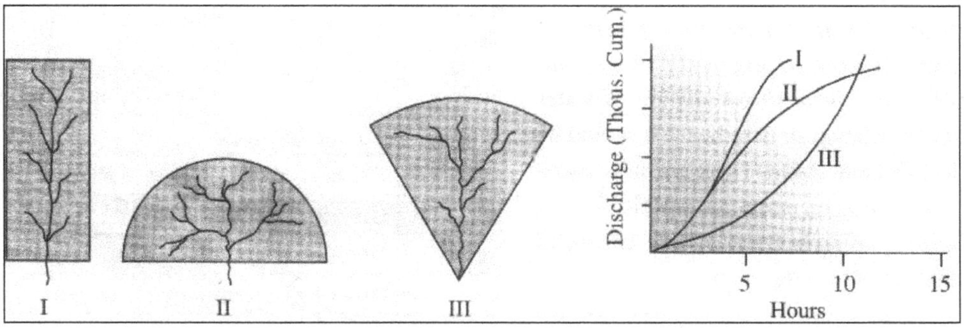

Changing rate of water ingress with time as per the shape of the catchment.

Change in Ecosystem

A riverine or fluvial system changes to lacustrine system leading to submergence of grassland and forestland causing ecological changes and alteration in biological productivity of various ecosystems. The traditional migration path of animals also gets submerged.

Relocation and Rehabilitation of Human Beings

In many of the cases, submergence of agricultural fields and human habitation takes place. This not only leads to displacement of populace but also loss of livelihood. At the very outset i.e. at planning stage, relocation, rehabilitation and provision of proper livelihood has to be on card for probable displaced populace and need to be executed before starting the project.

Leakage and Seepage of Reservoir Water

Another pertinent issue is leakage and seepage of water from reservoir which not only undermines the very purpose of reservoir but also cause excessive pore water pressure causing uplift below the dam foundation. Leakage is defined as water loss along the linear openings as stream threads and seepage of water is a loss as diffuse flow through interconnected pores. The causes of leakage can be ascribed to presence of fault and shear zones, high incidence of joints in rocks etc. while seepage may take place along alluvial fill, through the dam and through the foundation material. Another consequence of leakage and seepage of water is groundwater level rise in rocks surrounding the reservoir, which may add to the problem of land sliding. The only positive effect of this rise termed as bank storage, is its discharge in reservoir during lean period when reservoir water level is well below the requirement.

The loss of water from reservoir area can be estimated by Packer's Test wherein water is pumped in to a borehole and loss are measured in terms of Lugeon Unit. One Lugeon unit is defined as loss of water in 10 minutes through a one-meter section of a borehole under 1Ma of pressure. One Lugeon unit equals the coefficient of permeability of 10-7 m/sec and a rock having permeability less than this is taken to be practically impervious or water tight. It has been found that Lugeon values correspond very well with the opening of joint planes.

The protection against leakage and seepage is adopted depending upon worth of investment and value of water going as loss. In case of small and medium size, earth and rock fill dams impervious lining using natural and artificial material are used to prevent water leakage and seepage. If loss of water either undermines the purpose of the reservoir or is deleterious for the foundation of dam, then grouting is the only option as an effective measure.

Table: Examples of Some Typical Lugeon Values.

Lugeon Value	Strong Rocks with Continuous Jointing	Weak Heavily Jointed Rocks
0	Completely tight.	Completely tight.
1	Occasionally open up to 1 mm.	Occasionally open hair cracks < 0.3 mm.
2	Opening 1 to 2 mm.	Opening 0.3 to < 1 mm.
3 to 5	Opening occasionally up to 2.5 mm.	Opening occasionally up to 1 mm.
20	Opening Frequently up to 1 mm.	Opening occasionally up to 1 mm.
50	Opening Frequently up to 2.5 mm.	Opening frequently up to 2.5 mm.
100	Opening Frequently up to 6.0 mm.	Opening frequently up to 6.0 mm.

Sedimentation in Reservoir

The damming of river will automatically stop the river sediment in reservoir, which were ought to go downstream and get distributed all along its course. Though it is very rare to have rate of sedimentation in reservoir as a decisive factor in deciding the location of the dam but, is a profound factor in deciding the dead storage capacity, as the life of a reservoir is based on this space, which is eventually got to be filled by silt but in designed life of the dam. The decision for height of the dam may also have to take the rate of sedimentation into account, as the total reservoir capacity minus the dead storage will control the total water available for the dam is intended for.

Table: Rate of Siltation in some Indian Reservoirs.

Reservoir	River	State	Assumed ha m/100km²/yr	Observed ha m/100km²/y	Annual Storage Capacity Loss (%)
Pong	Beas	HP	4.29	15.10	0.13
Bhakra	Sutluj	Punjab	4.29	6.14	0.65
Kalagarh	Ramganga	Uttarakhand	4.29	17.30	0.25
Mayurkashi	Mahanadi	West Bengal	3.61	20.09	0.50
Matatila	Betwa	UP	1.43	3.50	0.80
Maithon	Barakar	Jharkhand	1.62	13.02	0.50

The cost of sluicing or dredging of sedimentation may be very costly to undertake and may ultimately result in abandonment of the dam-reservoir. Construction of check dams and small run off river dams along the contributing streams and vegetation turfing in catchment area are the best options to mitigate this problem.

Reservoir Induced Seismicity

If not all but few areas after the construction of dam-reservoir have shown spurt in seismicity termed as Reservoir Induced Seismicity (RIS). The excessive load due to rise in water level on rocks, devoid of such pressure prior to construction of dam-reservoir, start failing, causing seismicity. Shear zones and faults which are seemingly dead and locked at present state of the residual stresses in rock mass may get reactivated due to added load of water as well as seepage of water along them leading to slippage and occurrence of earthquakes.

Size and shape of the reservoir plays an important role in causing RIS. The fact that larger is the reservoir in terms of area, larger will be the water spread and lesser will be the load exerted by water column on the per unit area of bed rocks as compared to linear reservoirs where thick water column may exert high magnitude loads on bed rocks, an important factor for reservoir induced seismicity. Other important factors include:

- Maximum height of the reservoir water level.
- Rate of increase of water level.
- Duration to which high water level maintained.
- Deeper infiltration of impounded water through fractures, joints, shear zones, exposed and concealed faults or fault zones.
- Rocks with chances of undergoing brittle deformation.

Geotechnical Investigations for Tunnel

Tunnels are linear opening below the ground or through natural obstacles and act as an element of transport. The other horizontal variants of tunnels include water pressure tunnel, used for carrying water from reservoir to power generating unit, drift is smaller version of tunnel can be used as starting point for larger diameter tunnel, adit, a linear small opening as an approach to the tunnel for carrying out forward exploration and to act as conduit for men and machine. Half tunnel to act as underground by pass, Chunnelor subaqueous tunnel or bridge. Most of the above-mentioned linear openings are straight to curvilinear and horizontal to slightly sloping. Its vertical variants such as shaft and raise are used for taking men and machine in underground mines and also to provide ventilation for all kinds of underground openings. Tunnels and shafts are important part of all variants of underground mining.

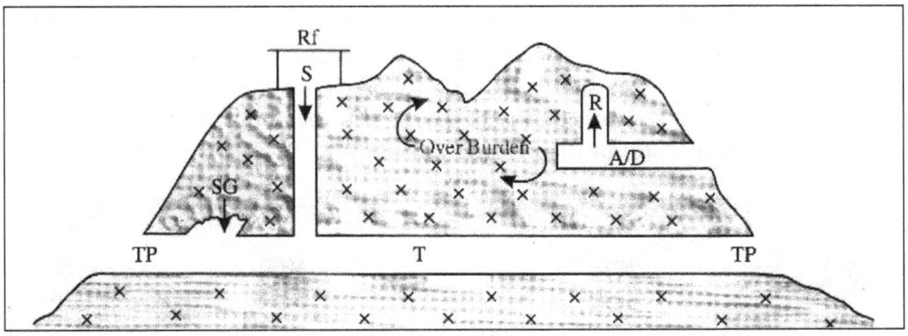

Different kinds of linear openings through a natural obstacle. Horizontal linear two end opening, tunnel portals (TP), one side opening, adit or drift (A/D). Vertical double end opening, shaft (S) with roof cover (Rf), and one end opening raise (R). Irregular opening, stope (SG), in a massive rock.

Underground power stations, railway stations, waste disposal sites (nuclear waste), bunkers and other strategic storage structures, involving excavation of rectangular openings termed as caverns also require to have geological and geotechnical explorations similar to tunnels.

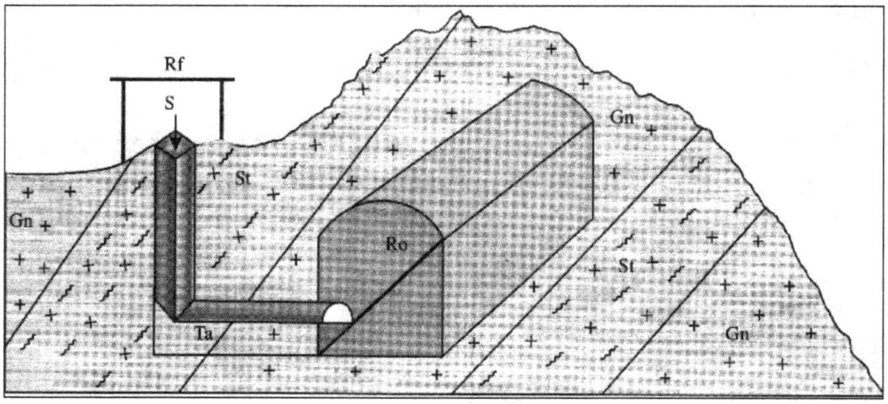

Rectangular opening with approach shaft (S) with roof (Rf), and tunnel (Ta) in inclined metamorphic rocks (Gn-Gneiss, St-Schist).

The alignment of a tunnel, probable support system, method of excavation, type of explosives, problem of groundwater, workability and disposal of rock waste, all need detailed geological investigation before starting the tunnel excavation or construction. The cross sectional and linear dimensions of tunnels will depend upon natural obstacles they have to go through and purpose of the tunnel and accordingly will be subjected to different kinds of geological problems.

Tunnelling Methods

Cut and Cover Tunnel

A method of tunnel construction involving the installation of earth support systems (e.g. slurry walls) followed by the main excavation, placing of the base slab, roof slab and subsequent backfilling to the final ground level.

Sequential Excavation Method

A method of tunnel construction that involves the use of standard construction equipment for excavation. The tunnel is usually lined in two steps:

- An initial lining of sprayed concrete provides immediate support and a subsequent secondary or permanent lining is then placed using either sprayed concrete or cast insitu concrete.
- A waterproof membrane is usually installed between the primary and secondary linings.

Slurry Wall

A form of earth support system whereby a continuous trench is excavated in the ground to the required depth, using slurry to provide temporary support during excavation of the trench. Reinforcement (which may be reinforcing cages or steel H sections) is lowered into the trench and concrete is subsequently placed by tremie pipe, displacing the slurry from the trench.

Tunnel Boring Machine

A method of tunnel construction that involves the procurement of a custom made piece of construction equipment. The TBM is equipped with a cutter head that is used to mine the ground. The excavation is continuously supported by installing precast concrete segments within the TBM and grouting them in place as the machine advances.

Road Tunnels

Road tunnels are feasible alternatives to cross a water body or traverse through physical barriers such as mountains. Road tunnels are viable means to minimize potential environmental impact such as traffic congestion, pedestrian movement, air quality, noise pollution, or visual intrusion; to protect areas of special cultural or historical value such as conservation of districts, buildings or private properties. Planning for a road tunnel requires multi-disciplinary involvement and assessments. Certain considerations, such as lighting, ventilation, life safety, operation and maintenance, etc. should be addressed specifically for tunnels.

Tunnelling Route Studies

In a tunnel route study, the following issues should be considered:

- Subsurface, geological and geo-hydraulic conditions
- Constructability
- Long-term environmental impact
- Seismicity
- Land use restrictions
- Potential air right developments
- Life expectancy
- Economical benefits and life cycle cost
- Operation and maintenance
- Security
- Sustainability

Geotechnical Investigations for Tunnelling

- Geotechnical investigations are critical for proper planning of a tunnel.
- Selection of the alignment, cross section, and construction methods is influenced by the geological and geotechnical conditions, as well as the site constraints. Good knowledge of the expected geological conditions is essential. Tunnel alignment is sometimes changed based on the results of the geotechnical to minimize construction cost or to reduce risks.

- The type of the ground encountered along the alignment would affect the selection of the tunnel type and its method of construction.
- Study of the impact of geological features on the tunnel alignment in the presence of active or inactive faults. During the planning phase, avoid crossing a fault zone. If it is unavoidable then proper measures for crossing it should be implemented. Presence of faults or potentially liquefiable materials would be of concern during the planning process.
- Geotechnical issues such as the soil or rock properties, the ground water regime and the ground cover over the tunnel should be analysed. The investigation should address not just the soil and rock properties, but also their anticipated behaviours during excavation.
- The investigation should also address groundwater. For example, in soft ground SEM tunneling, the stability of the excavated face is greatly dependent on control of the groundwater. Dewatering, pre-draining, grouting, or freezing are often used to stabilize the excavation.
- Analysing the ground behavior during tunneling will affect potential settlements on the surface. Measures to minimize settlements by using suitable tunneling methods or by preconditioning the ground to improve its characteristics would be required.
- Risk assessment is an important factor in selecting a tunnel alignment. Construction risks. Sensitive existing structures. Very Hard spots (rock, for example) beneath parts of a tunnel.

Tunnel Type Studies

General Description of Various Tunnel Types: The principal types and methods of tunnel construction that are in use are:

- Cut-and-cover tunnels are built by excavating a trench, constructing the concrete structure in the trench, and covering it with soil. The tunnels may be constructed in place or by using precast sections
- Bored or mined tunnels, built without excavating the ground surface. These tunnels are usually labelled according to the type of material being excavated. Sometimes a tunnel passes through the boundary between different types of material; this often results in a difficult construction known as mixed face.
- Rock tunnels are excavated through the rock by drilled and blasting, by mechanized excavators in softer rock, or by using rock tunnel boring machines (TBM). In certain conditions, Sequential Excavation Method (SEM) is used.
- Soft ground tunnels are excavated in soil using a shield or pressurized face TBM.
- Principally earth pressure balance or slurry types, or by mining methods, known as either the sequential excavation method (SEM).
- Immersed tunnels are made from very large precast concrete or concrete-filled steel elements that are fabricated in the dry, floated to the site, placed in a prepared trench below water, and connected to the previous elements, and then covered up with backfill.

- Jacked box tunnels are prefabricated box structures jacked horizontally through the soil using methods to reduce surface friction; jacked tunnels are often used where they are very shallow but the surface must not be disturbed, for example beneath runways or railroads embankments.

The selection of a tunnel type depends on the geometrical configurations, the ground conditions, the type of crossing, and environmental requirements.

Design Process

The basic process used in the design of a road tunnel is:

- Define the functional requirements, including design life and durability requirements.
- Carry out the necessary investigations and analyses of the geologic, geotechnical and geo-hydrological data.
- Conduct environmental, cultural, and institutional studies to assess how they impact the design and construction of the tunnel.
- Perform tunnel type studies to determine the most appropriate method of tunneling.
- Establish design criteria and perform the design of the various tunnel elements.
- Appropriate initial and final ground support and lining systems are critical for the tunnel design, considering both ground conditions and the proposed method of construction.
- Perform the design in Preliminary and Final design phases. Interim reviews should be made if indicated by on-going design issues.
- Establish tunnel alignment, profile and cross-section.
- Determine potential modes of failure, including construction events, unsatisfactory long-term performance, and failure to meet environmental requirements. Obtain any necessary data and analyse these modes of failure.
- Perform risk analysis and identify mitigation measures and implement those measures in the design.
- Prepare project documents including construction plans, specifications, schedules, estimates, and geotechnical baseline report (GBR).

Tunnel Cross-Section

The tunnel cross section geometrical configuration must satisfy the required traffic lanes, shoulders or safety walks, suitable spaces for ventilation, lights, traffic control system, fire/life safety systems, etc. The cross section is also dictated by the method of tunnel construction.

The available spaces in a circular cross section can be used to house tunnel systems, such as the ventilation duct or fans, lighting, traffic control systems and signs, close circuit TV, and the like. For rectangular sections the various systems can be placed overhead, invert or adjacent to the traffic lanes if overhead space is limited.

The tunnel structural systems depend on the type of tunnel, the geometrical configuration of the cross section, and method of construction. In addition no dripping or visible leakage from a single location shall be permitted.

Tunnel waterproofing systems are used to prevent groundwater inflow into an underground opening. They consist of a combination of various materials and elements. The design of a waterproofing system is based on the understanding of the ground and geohydrological conditions, geometry and layout of the structure and construction methods to be used.

A waterproofing system should always be an integrated system that takes into account intermediate construction stages, final conditions of structures and their ultimate usage including maintenance and operations. There are two basic types of waterproofing systems: drained (open) and undrained (closed).

Various waterproofing materials are available for these systems. Open waterproofing systems allow groundwater inflow into a tunnel drainage system. Groundwater inflow is typically localized to distinct locations such as joints and fractures and the overall permeability is such that a groundwater draw-down in soil layers overlying the rock mass will not be affected.

Horizontal And Vertical Alignments

Planning and design of road tunnel alignments must consider the geological, geotechnical and groundwater conditions at the site.

Geotechnical Investigations

To successfully plan, design and construct a road tunnel project requires various types of investigative techniques to obtain a broad spectrum of pertinent topographic, geologic, subsurface, geo-hydrological, and structure information and data.

Although most of the techniques and procedures are similar to those applied for roadway and bridge projects, the specific scope, objectives and focuses of the investigations are considerably different for tunnel and underground projects, and can vary significantly with subsurface conditions and tunneling methods.

A geotechnical investigation program for a tunnel project must use appropriate means and methods to obtain necessary characteristics and properties as basis for planning, design and construction of the tunnel and related underground facilities, to identify the potential construction risks, and to establish realistic cost estimate and schedule.

The extent of the investigation should be consistent with the project scope (i.e., location, size, and budget), the project objectives (i.e., risk tolerance, long-term performance), and the project constraints (i.e., geometry, constructability, third-party impacts, aesthetics, and environmental impact).

It is important that the involved parties have a common understanding of the geotechnical basis for design, and that they are aware of the inevitable risk of not being able to completely define existing subsurface conditions or to fully predict ground behavior during construction.

Generally, an investigation program for planning and design of a road tunnel project may include the following components:

- Existing Information Collection and Study
- Surveys and Site Reconnaissance
- Geologic Mapping
- Subsurface Investigations
- Environmental Studies
- Seismicity
- Geospatial Data Management

Typical stages of a road tunnel project from conception to completion are:

- Planning
- Feasibility Study
- Corridor and Alignment Alternative Study
- Environmental Impact Studies (EIS) and Conceptual Design
- Preliminary Design
- Final Design

The early investigations for planning and feasibility studies can be confined to information studies and preliminary reconnaissance. Geological mapping and minimum subsurface investigations are typically required for EIS, alternative studies and conceptual design. EIS studies may also include limited topographical and environmental investigations. The subsurface investigation techniques typically used for planning, design and construction of road tunnels.

Information Study

The first phase of an investigation program for a road tunnel project starts with collection and review of available information to develop an overall understanding of the site conditions and constraints at little cost.

Existing data can help identify existing conditions and features that may impact the design and construction of the proposed tunnel, and can guide in planning the scope and details of the subsurface investigation program to address these issues.

Published topographical, hydrological, geological, geotechnical, environmental, zoning, and other information should be collected, organized and evaluated. In areas where seismic condition may govern or influence the project, historical seismic records are used to assess earthquake hazards.

Records of landslides caused by earthquakes, documented by the USGS and some State Transportation Departments, can be useful to avoid locating tunnel portals and shafts at these potentially unstable areas.

Topographical Data

Topographic maps and aerial photographs that today can be easily and economically obtained, are useful in showing terrain and geologic features (i.e., faults, drainage channels, sinkholes, etc.). When overlapped with published geological maps they can often, by interpretation, show geologic structures. Aerial photographs taken on different dates may reveal the site history in terms of earthwork, erosion and scouring, past construction, etc.

Surveys and Site Reconnaissance

Site Reconnaissance and Preliminary Surveys

The lower-resolution contour maps are sufficient only for planning purposes. However, a preliminary survey will be needed for concept development and preliminary design to expand existing topographical data and include data from field surveys and an initial site reconnaissance. Initial on-site studies should start with a careful reconnaissance over the tunnel alignment, paying particular attention to the potential portal and shaft locations.

Features identified on maps and air photos should be verified. Rock outcrops, often exposed in highway and railroad cuts, provide a source for information about rock mass fracturing and bedding and the location of rock type boundaries, faults, dikes, and other geologic features. Features identified during the site reconnaissance should be photographed, documented and if feasible located by hand-held GPS equipment.

Sources of Information Data (After FHWA, 2002a)

Source Functional Use Location Examples: Local Soil Conservation.

Aerial Photographs

- Identifies manmade structures.
- Provides geologic and hydrological information can be used as a basis for site reconnaissance.
- Track site changes over time.
- Topographic Maps.
- Geologic Maps and Reports.
- Prior Subsurface Investigation Reports.
- Prior Underground and Foundation details.
- Construction Records.

- Provides information on local soil/rock type.
- Strength parameters.
- Hydrogeological issues.
- Environmental concerns.
- Tunnel construction methods and problems.

Water Well Logs

- Provide stratigraphy of the site and/or regional areas.
- Yield rate and permeability, Groundwater levels.

Control Survey

The reconnaissance should cover the immediate project vicinity, as well as a larger regional area so that regional geologic, hydrologic and seismic influences can be accounted for. A preliminary horizontal and vertical control survey may be required to obtain general site data for route selection and for design.

Topographic Surveys

For tunnel projects, detailed topographic maps, plans and profiles must be developed to establish primary control for final design and construction based on a high order horizontal and vertical control field survey.

On a road tunnel system, center-line of the roadway and center-line of tunnel are normally not identical because of clearance requirements for walkways and emergency passages. A tunnel center line developed during design should be composed of tangent, circular, and transition spiral sections that approximate the complex theoretical tunnel center line within a specified tolerance (0.25 in.).

Accurate topographic mapping is required to support surface geology mapping and the layout of exploratory borings, whether existing or performed for the project. The principal survey techniques include:

- Conventional Survey
- Global Positioning System (GPS)
- Electronic Distance Measuring (EDM) with Total Stations
- Remote Sensing
- Laser Scanning

Global Positioning System (GPS) helps to determine the relative position of monuments in a control network. GPS surveying is able to coordinate widely spaced control monuments for long range surveys, as well as shorter range surveys. A Total Station is also used for this purpose.

Remote Sensing can effectively identify terrain conditions, geologic formations, escarpments and surface reflection of faults, buried stream beds, site access conditions and general soil and rock formations.

Hydrographical Surveys

Hydrographic surveys are required for subaqueous tunnels including immersed tunnel, shallow bored tunnel, jacked box tunnel, and cofferdam cut-and-cover river crossings to determine bottom topography of the water body, together with water flow direction and velocity, range in water level, and potential scour depth.

In planning the hydrographic survey, an investigation should be made to determine the existence and location of submarine pipelines, cables, natural and sunken obstructions, rip rap, etc. that may impact design or construction of the immersed tunnel or cofferdam cut-and-cover tunnel.

Additional surveys such as magnetometer, seismic sub-bottom scanning, electromagnetic survey, side scan sonar, etc., may be required to detect and locate these features. These additional surveys may be done simultaneously or sequentially with the basic hydrographic survey.

Geologic Mapping

After collecting and reviewing existing geologic maps, aerial photos, references, and the results of a preliminary site reconnaissance, surface geologic mapping of available rock outcrops should be performed by an experienced engineering geologist to obtain detailed, site-specific information on rock quality and structure.

Geologic mapping collects local, detailed geologic data systematically, and is used to characterize and document the condition of rock mass or outcrop for rock mass classification, such as:

- Discontinuity type
- Discontinuity orientation
- Discontinuity infilling
- Discontinuity spacing
- Discontinuity persistence
- Weathering

In addition, the following surface features should also be observed and documented during the geologic mapping program:

- Slides, new or old, particularly in proposed portal and shaft areas
- Faults
- Rock weathering

- Sinkholes and karstic terrain
- Groundwater springs
- Volcanic activity
- Anhydrite, gypsum, pyrite, or swelling shales
- Stress relief cracks
- Presence of talus or boulders
- Thermal water (heat) and gas

The mapping data will also help in targeting subsurface investigation borings and in situ testing in areas of observed variability and anomalies.

Subsurface Investigations

Ground conditions including geological, geotechnical, and hydrological conditions, have a major impact on the planning, design, construction and cost of a road tunnel, and often determine its feasibility and final route.

Fundamentally, subsurface investigation is the most important type of investigations to obtain ground conditions, as it is the principal means for:

- Defining the subsurface profile (i.e. stratigraphy, structure, and principal soil and rock types).
- Determining soil and rock material properties and mass characteristics.
- Identify geological anomalies, fault zones and other hazards (squeezing soils, methane gas, etc.)
- Defining hydrogeological conditions (groundwater levels, aquifers, hydrostatic pressures, etc.).
- Identifying potential construction risks (boulders, etc.).

Subsurface investigations typically consist of borings, sampling, in situ testing, geophysical investigations, and laboratory material testing.

The principal purposes of these investigation techniques are summarized below:

- Borings are used to identify the subsurface stratigraphy, and to obtain disturbed and undisturbed samples for visual classification and laboratory testing.
- In situ tests are commonly used to obtain useful engineering and index properties by testing the material in place to avoid the disturbance inevitably caused by sampling, transportation and handling of samples retrieved from boreholes; in situ tests can also aid in defining stratigraphy.

- Geophysical tests quickly and economically obtain subsurface information (stratigraphy and general engineering characteristics) over a large area to help define stratigraphy and to identify appropriate locations for performing borings.
- Laboratory testing provides a wide variety of engineering properties and index properties from representative soil samples and rock core retrieved from the borings.

Test Borings and Sampling

Various field testing techniques can be performed in conjunction with the test borings as well. In general, borings should be extended to at least 1.5 tunnel diameters below the proposed tunnel invert. Borings at shafts should extend at least 1.5 times the depth of the shaft for design of the shoring system and shaft foundation, especially in soft soils.

Sampling – Overburden Soil

Standard split spoon (disturbed) soil samples (ASTM D-1586) are typically obtained at intervals not greater than 5 feet and at changes in strata. Continuous sampling from one diameter above the tunnel crown to one diameter below the tunnel invert is advised to better define the stratification and materials within this zone if within soil or intermediate geo-material. In addition, undisturbed tube samples should be obtained in each cohesive soil stratum encountered in the borings; where a thick stratum of cohesive soil is present, undisturbed samples should be obtained at intervals not exceeding 5m.

Sampling – Rock Core

In rock, continuous rock core should be obtained below the surface of rock, with a minimum core diameter of 2.16 inch or 54.7 mm. For deeper holes, coring should be performed with the use of wire-line drilling equipment to further reduce potential degradation of the recovered core samples. The rock should be logged soon after it was extracted from the core barrel.

Borehole Sealing

All borings should be properly sealed at the completion of the field exploration, if not intended to be used as monitoring wells. This is typically required for safety considerations and to prevent cross contamination of soil strata and groundwater.

Test Pits

Test pits are often used to investigate the shallow presence, location and depth of existing utilities, structure foundations, top of bedrock and other underground features that may interfere or be impacted by the construction of shafts, portals and cut-and-cover tunnels. The depth and size of test pits will be dictated by the depth and extent of the feature being exposed.

Geophysical Testing

Geophysical tests are indirect methods of exploration in which changes in certain physical characteristics such as magnetism, density, electrical resistivity, elasticity, or a combination of these are used as an aid in developing subsurface information. Geophysical methods provide an expeditious and economical means of supplementing information obtained by direct exploratory methods, such as borings, test pits and in situ testing; identifying local anomalies that might not be identified by other methods of exploration and defining strata boundaries between widely spaced borings for more realistic prediction of subsurface profiles.

Typical uses of geophysical tests include determination of the top of bedrock, the rip ability of rock, the depth to groundwater, the limits of organic deposits, the presence of voids, the location and depth of utilities, the location and depth of existing foundations, and the location and depth of other obstruction, to note just a few.

Groundwater Investigation

Groundwater is a major factor for all types of projects, but for tunnels groundwater is a particularly critical issue since it may not only represent a large percentage of the loading on the final tunnel lining, but also it largely determines ground behavior and stability for soft ground tunnels; the inflow into rock tunnels; the method and equipment selected for tunnel construction; and the long-term performance of the completed structure.

Groundwater investigations typically include most or all of the following elements:

- Observation of groundwater levels in boreholes.
- Assessment of soil moisture changes in the boreholes.
- Groundwater sampling for environmental testing.
- Installation of groundwater observation wells and piezometers.
- Borehole permeability tests (rising, falling and constant head tests; packer tests, etc.)
- Geophysical testing.
- Pumping tests.

Environmental Issues

Although tunnels are generally considered environmentally-friendly structures, certain short-term environmental impacts during construction are unavoidable. Long-term impacts from the tunnel itself, and from portals, vent shafts and approaches on local communities, historic sites, wetlands, and other aesthetically, environmentally, and ecologically sensitive areas must be identified and investigated thoroughly during the project planning and feasibility stages, and appropriately addressed in environmental studies and design.

Seismicity

The release of energy from earthquakes sends seismic acceleration waves traveling through the ground. The factors that can affect the response of the ground during earthquakes:

- Distance of the seismic source from the project site.

- Magnitude of the seismic accelerations.
- Earthquake duration.
- Subsurface profile.
- Dynamic characteristics and strengths of the materials affected.

Geological Considerations for Construction

Road and Railway

Road and rail networks and bridges are the lifeline infrastructures facilities for any region or country. Roads being the oldest and most important element of transport have played a pivotal role in the evolution of civilizations. With the beginning of Industrial era and the advent of steam engine, railways have supplemented the transportation of people and goods in a big way and are still vital for modern day industrial growth. The bridges being the integral part of the road and rail networks spanning not only over rivers but also over arms of sea to provide all season connectivity. The understanding of topography, geomorphology and geology of the area being traversed by these transportation elements is important for safe design, longevity and economy. As huge amount of construction material in form of earth and rocks will be required in laying of roads, rails and construction of bridges. Hence, their availability must also be ensured keeping in view its quality and haulage cost.

Road and Rail Network

From time immemorial, the construction material and methods have developed from stone pavements in the beginning during ancient period, through kankar and limesurkhi-pitch roads in medieval period to modern day tar and concrete roads with state of the art construction technology. Nowadays the quality and connectivity of road network not only forms the basis but also is one of the indices to show infrastructural development of a country. There are two types of roads termed as Flexible Pavement and Rigid Pavement. Flexible pavement, is said so because of the fact that part of road may flex down or raised up due to differential compaction of founding soil or rocks. It is laid on original founding soil called as sub grade, followed by earth used for raising the level termed as sub base course, followed by different layers of rock aggregate as base course, surface course and wearing course. In the upper most layer, termed as wearing course, coarse sand or rock powder of same size is sprayed over to fill the small pores with the help of tar. Flexible pavements are the most versatile and prevalent road type, all over the world. The rigid pavement gets its name from the cast concrete (CC) or reinforced concrete layer (RCC), founded over soil or rock foundation and sub base made of rock aggregate. The CC or RCC layer acts as slab, does not allow any settlement, and normally constructed in areas having problem of water logging. The flexible pavements are cheaper and can be maintained at low cost while CC and RCC roads may be more durable but involves high construction and high maintenance cost.

The railways are also laid on raised ground over the founding ground with borrowed earth compacted at optimum moisture condition. It is then covered by rail ballast of 40-80 mm size for 40 to 60 cm thick apron followed by 40 to 50 mm ballast for shoulder of rail tracks. Highways and railways both usually pass through different kinds of terrains and varied founding substrate. In plains, it is usually or alluvial or at the most Aeolian soil deposits with very gentle slopes. In plateaux topography, they may come across with both earthy and rocky ground with low to moderate angle slopes. However, when it comes to hilly areas it has to negotiate high angle slopes over different types of rocks with differing attitude. Especially in case of railway tracks, sharp bends and sudden slope change cannot be allowed as against roads, which rise on slopes with sharp hairpin bends. In this way, railway tracks need more space sideways and also more number of bridges as compared to roads. The long-term stability of roads and railways is of prime importance especially on rocky slopes whose failure can have serious consequences. In the following section problems faced by roads and rail-tracks in hills are discussed.

Route Alignment

The route alignment between destination to destination is the most important factor as it will not only decide the length of the route but also the slope it has to negotiate, bridges and tunnels to be made and rock sequences it has to come across. Normally in case of laying new road or railways, many different alignments have to be studied thoroughly by taking into account all above factors, logistics involved in terms of availability of construction material, machine and economy. The important investigations involved along highway and railways include following:

- The total running distance is studied to assess geotechnical properties for deciding the alignment, to identify problematic areas, and to decide its mending or bypassing.
- The volume of cutting and filling required, identification of borrowed areas for getting earth and rocks.
- Assessment of instability of slopes along the alignment and measures to be taken for its stability.
- Identification and of organics rich soil, swamps, palaeo-channels, cavities and peat deposits to ascertain there effect on stability.
- Making of detailed geological maps incorporating the road and its sides showing different types of soil, thickness of soil cover, rocks and its attitude.

Remote sensing imageries, aerial photographs, Survey of India topographic maps and existing geological maps are used to decide different possible alignments, identification of potential areas of land sliding, number of rivers to come across and to mark major topographic obstacles to decide construction of tunnels along the route. The ground study and surveys are then taken up to decide final one, from the above alignments.

Factors Influencing Stability of Roads and Railways

The stability of a particular rock slope is governed by natural geomorphic and geological factors as well as on excavation methods employed during construction. The most important geomorphic factor is nature of slope, while important geologic factors include lithology, its structure and weathering grade. The construction of roads invariably involves excavations of rocky slopes to make benches for laying roads. As slopes created for roads will be subjected to various kinds of mass movements, their stability will be controlled by:

- Type of rocks, its strength, density, cohesion.
- Alignment of cut slope and bench with respect to attitude of bedding, foliation.
- Angle of slope and ratio of slope height with that of bench width termed as batter.
- Presence and orientation of joints, shear zones, fault and fault zones.
- Flow pattern of surface and ground water.

Rock Types

The overall behavior of rock mass will depend on its intrinsic properties such as density, porosity, strength, cohesion a factor of mineral composition and texture of rocks as well as on the presence of joints and shear zones in a rock mass. Igneous rocks and non-foliated metamorphic rocks with interlocking texture with homogenous mineralogy will be the most favorable rocks. The roads in regions with sedimentary and foliated metamorphic rocks are more likely to suffer by landslides as they are intrinsically layered.

Ultramafic igneous rocks pyroxinite, serpentine; sedimentary rocks such as evaporites, marl, shale, chalk and metamorphic rocks such as schist, phyllite and slate are to be specially watched if they come along the route alignment as they are known weak rocks. The rocks are subjected to change by weathering specially if rocks are geologically very old, lying in regions with humid tropical climate, will be prone to landslides for example railroads in Konkan area. Similarly, roads and railways in geo-tectonically active areas (Himalayan Mountain) are subjected to landslides due to high angle slopes, sudden heavy downpours and earthquakes, as in Uttarakhand and Himachal Pradesh.

Attitude of Rocks

The strike and dip of layers in rocks such as volcanic flows, bedding, lamination and foliation planes with respect to slope cut and bench plays a very important role. Horizontal rocks will provide moderate to good conditions while vertical rocks will pose problems on cut slope and bench. The most favourable situation is when rocks dipping against the slope, while most unfavourable condition will be when rocks dip towards slope. If the hills have horizontal or vertical rocks, the slope created will have almost similar conditions all around the hill. The dipping rocks will have different conditions around the hill i.e. in hill face perpendicular to strike of rocks will dip towards the slope, opposite to this face rocks will dip against the slope while in rest of the sections, rocks will show varying angles of dip (apparent dips) and may look apparently horizontal on slopes running parallel to its strike. In such cases it is prudent to have roads on the slope wherein rocks dip opposite to slope by having hair pin bends as against on slopes with rocks dipping towards it.

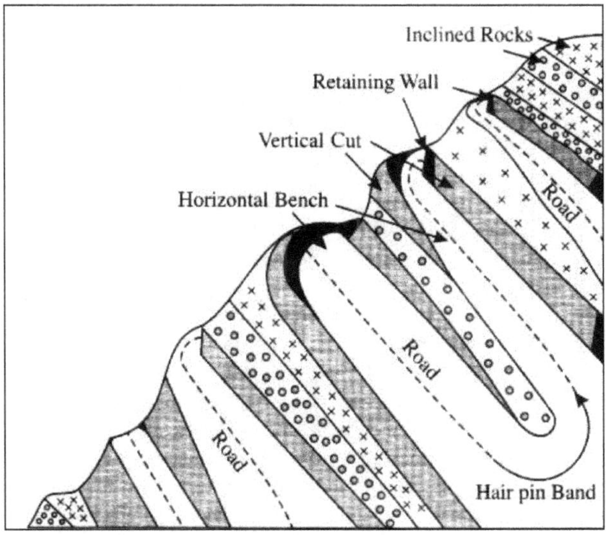

Road on a hill with rocks dipping opposite to slope, forming hairpin bands. Also, see the vertical and horizontal cut to form bench for laying road.

Batter and Slope Angle

Construction of roads on slope requires creation of benches to seat it. The width of bench will depend upon the proposed width of road and associated elements such as side slope, provision of drain, parapet and retaining wall etc. The ratio of vertical cut height with that of horizontal bench width termed as batter is very important for stability of the roads, which depends upon the rock type and its overall strength.

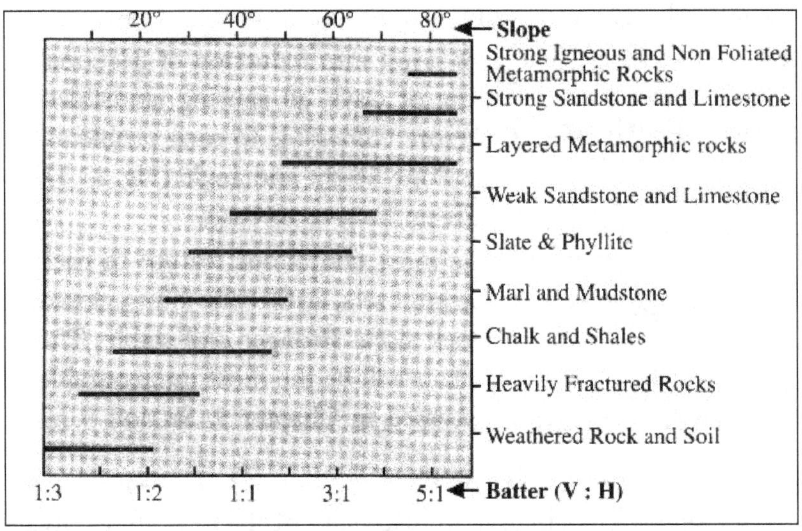

The ratio of vertical to horizontal cut to create road space on hill slopes for different rocks.

The inclination and height of side slope are also important as they are going to control the space, would be available after cut and fill for laying the road. Weathered and weak rocks will offer low height vertical cut and low angle slopes while strong rocks can afford to have high slopes with high angle of inclination. Steep and high slope cut economize the space hence lesser excavation. Sound rocks can be cut to almost vertical faces, raked back by 10° for safety and can be benched at every 5 to 10m height intervals.

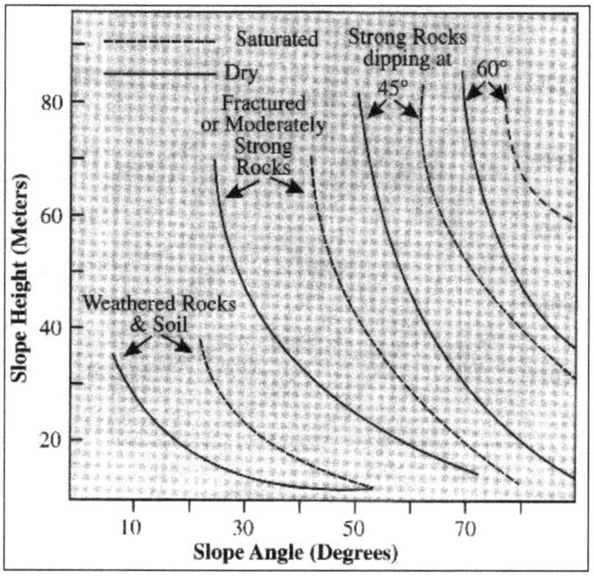

Rocks of different strength and attitude on different slope height, angle and saturation level.

Surface and Groundwater

Water is the most important factor in aiding and triggering the landslides. Surface runoff caused by rainfall result into sudden mass transportation of debris on the slopes. Infiltration of water through pores and fractures result in to reduction of frictional resistance along fractures, increased pore pressure and swelling of clays all causing landslides. Freshly exposed slopes and benches cut for making roads are more vulnerable to such problems. The presence or absence of groundwater in rocks too influence slope height and angle. Proper and improved drainage can ensure the stability of roads on long-term basis.

Joints, Shear Zones, Fault and Fault Zone

Joints the most ubiquitous feature in rock mass will control the overall rock mass characteristics. Lesser the joint sets, farther the joints, closed joint apertures and high degree of asperities will make the rock slope stronger and stable. The joint sets dipping with 45 to 75° and day lighting on the slope face will be the most unfavourable and need to be taken into account.

Excavation to Create Bench and Slope for Motor and Railway

To create space for laying road or railways on the shoulder of hills we need to create benches having different batter based on the rock mass conditions. For this different excavation techniques can be employed depending upon the rock types. Direct excavation by face shovel or by backhoe can be used for weathered and weak rock mass. Ripping can be employed for moderately weathered rocks of class IV using tractor mounted rippers or by breaking rocks using pneumatic hydraulic pecker. The old and most common technique has been the blasting if rocks are fresh, less fractured and strong. The different methods of excavation can be related to different rock mass parameters such as Rock Quality Designation, Unconfined Compressive Strength, Point Load Strength and fracture spacing.

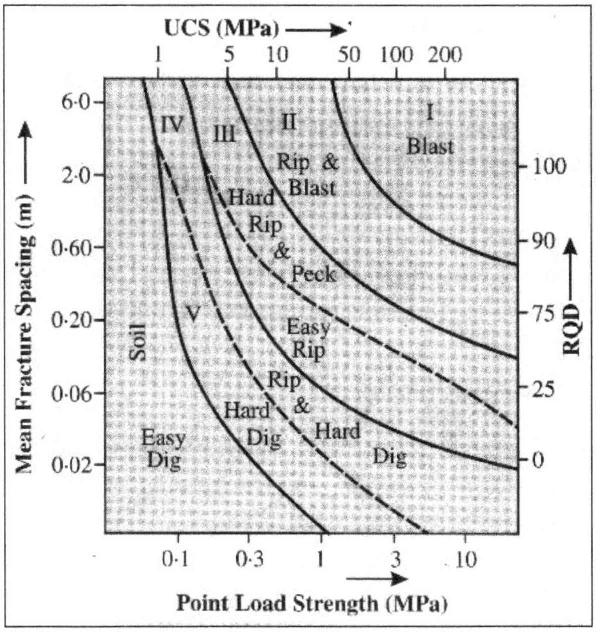

A graph to show different excavation method for different rock mass classes (I – V), based on various rock and rock mass parameters.

Open Face Blasting

A generalized schematics is shown in figure for blasting and excavation. It involves drilling of holes of 50 to 100mm diameter (D) usually raked at 10 to 15° of angle from the vertical. The rocks over the hole up to free face is called as burden (B) and ideally should be 30 to 40 times the D which may come to be 1.5 to 4.0m. Spacing between the drills holes are kept at 50 times the D, generally comes to be 3 to 5m. To make excavation smooth and to protect the slope from getting weak due to explosion few drill holes at a distance of 10 to 20 times D are kept with 10% less explosives that too decoupled i.e. not fully compacted.

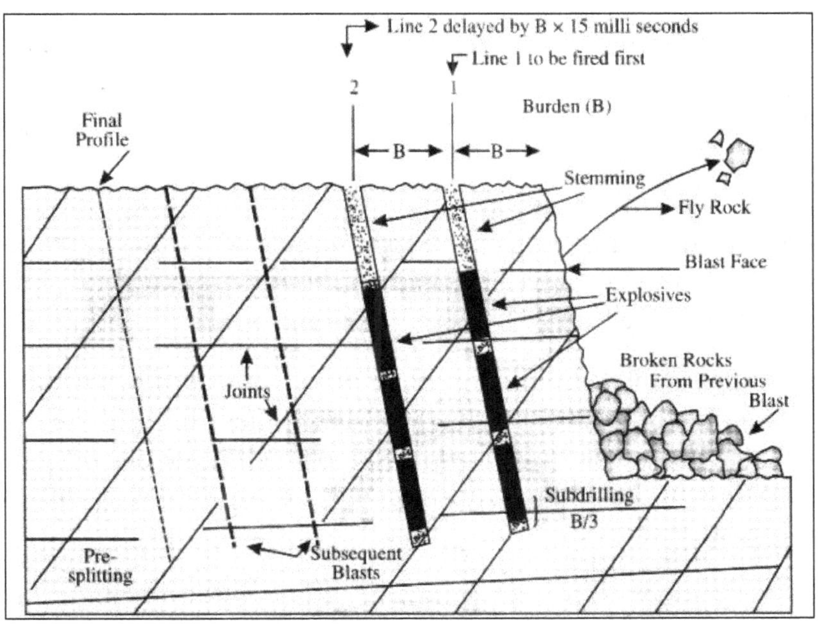

A pictorial representations of drilling and blasting for open face excavation.

Explosion Material and Method

Black Powder also known as Gun Powder composed of KNO_3 with S and C is used for low magnitude blasting. Nitro-glycerin with Ammonium Nitrate called as Dynamite is used for powerful explosion. Ammonium Nitrate (94%) with Fuel Oil (6%) termed as ANFO is cheap and safe to use for weak and fractured rock. The method involves charging where in explosive is kept in drill hole, is rammed tightly for good results the upper most part of the hole almost equal to burden, is left without explosive and is filled by sand termed as, stemming. Explosives are fired by using electric detonator by giving a delay, equal to 5 milliseconds/m of burden.

Engineering Geology of Bridge Site

Bridges are an integral part of road or rail track. In hilly areas, the number of bridges increases many times due to presence of numerous valleys with or without streams or rivers, as compared to their numbers in plain areas where they have to cross them occasionally. An engineering geologist ordinarily may not suggest the particular site for the bridge, as its location, height, span and type will be controlled by grade (slope) and alignment of the route and nature of traffic which are usually decided during the planning and designing stage. Though, he is very much involved in working out the geology of the bridge site, depth and condition of bed rock, thickness of overburden in river section, presence of buried channel, and correlation of rocks exposed on the two banks of the river where terminal support i.e. abutment of bridge will be placed.

Types of Bridges

A bridge consists of a superstructure placed on substructure, which is founded on a substrate comprising soil or soil and rock or only rock(s). The superstructure mainly consists of deck, placed on substructure either directly upon abutment or on last piers, which in turn are placed on foundation. There are four common types of bridges termed as:

- Girder Bridge,
- Cantilever Bridge,
- Arch Bridge,
- Suspension bridge.

Simple Beam or Girder Bridge is made up of one span supported by abutments at their ends. When more than one span is required, piers are used to provide support in middle parts. Now days beam bridges are replaced by Box Steel Girder bridges.

Cantilever or Truss Bridge is made up of more than one span with intermediate support of piers. Mainly Consist of assembly of triangles using steel bars where in rigid steel arms extend from both side of the piers and steel tubes projected from top and bottom of each piers hold the arm in place. The arms projecting in the mid part are supported at one end only called cantilever, which supports the central span.

Arch Bridge has its deck supported by arch action, in earlier days were made up of stones but now are made of steel or concrete.

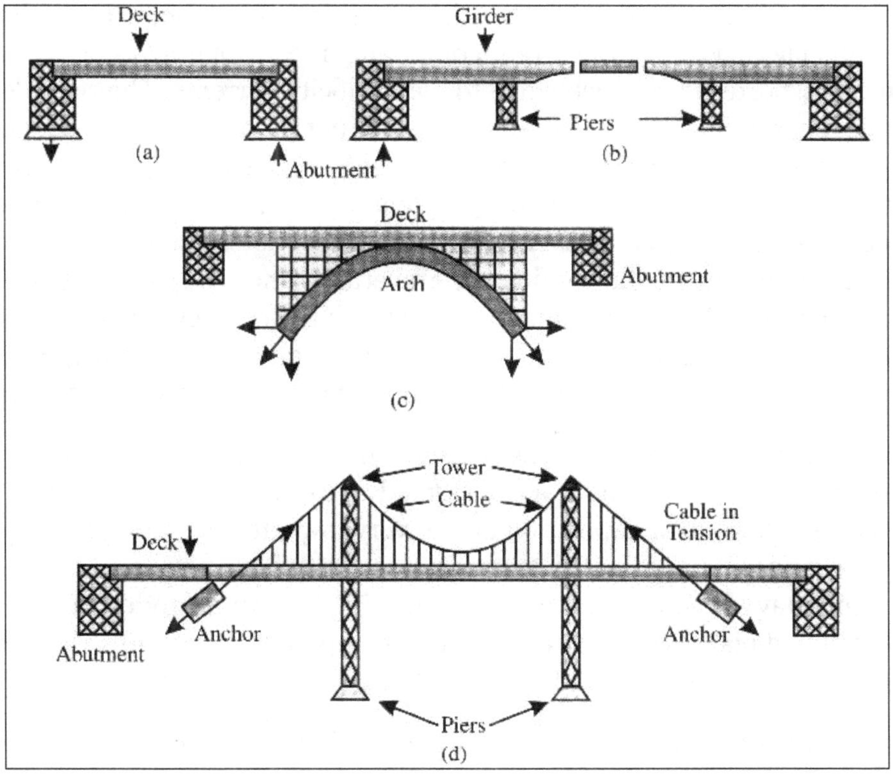

Figure shows Identify deck, abutments, piers and foundation part of bridges. (a) Simple beam or Girder bridge on end terminals/abutments; (b) Cantilever or truss bridges with end and mid pier supports; (c) Arch Bridge, see the orientation of forces; (d) Suspension bridge supported by cables anchored in rocks.

Suspension Bridge can have very large span i.e. 200 to 2000m largest of all bridge types consist of two strong steel cables, resting on saddles, fixed on steel or masonry towers and anchored in rocks at both ends, while deck is suspended on cables.

Geomorphological Factors

Any bridge on a river will affect the geomorphological process of a river; hence, basic knowledge of Fluvial Geomorphology is very important. The geomorphology of a river system can help in understanding the hydrodynamics and dynamic equilibrium of a river system at and around the bridge site. Normally a bridge negotiates a river at 90° to keep the length of bridge to its minimum. For this, the orientation of road has to be changed as it approaches the river. Depending upon the stages of river system the river may have different sizes in terms of its width, depth and cross section. Similarly, river may show different hydrodynamic character depending upon the gradient, discharge, nature and amount of sediment it is carrying with it. In addition, the order of a river system, catchment area, rainfall characteristics and drainage basin shall control how fast and how much of water comes into river in an event of sudden rainfall.

Fluvial Geomorphology

A river termed as Anastomosing River when it flows in hilly areas and is said to be in its Nascent

Stage, where rate of erosion is much faster than the deposition of sediment. The sediment deposit is usually thin channel lag deposit over bedrock, comprising mostly of gravels of all sizes with coarse grain sand being transported very quickly due to high velocity currents. The river depth is more as compared to its width and usually has 'V' shape unless it is not carved out by glaciers in which case 'U' shape valley is common. The river may have tortuous route in deeply carved out entrenched valleys. The rivers once come out of hilly tracts in adjoining plain are said to be in Young Stage, where erosion almost equals the deposition and rivers show straighter route. The valleys become very broad with thick sediment deposit of medium to fine size gravel and sand. The river may consist of multiple channels, called as Braided Rivers with intervening braid bars in valleys having breadth more than depth. The current velocity, nature and amount of sediment vary with amount of water during rainy and non-rainy season and at the same time; the location of main or active channel may also change.

The rivers in plain areas flow with tortuous route having meanders of different sizes and shapes and called as Meandering River. This stage is termed as Mature Stage where rate of erosion becomes much less than the rate of deposition due to very low current velocity, except during rainy season. The river changes its course due to erosion on outer bend and deposition on inner bend of river channel and may also make new course by leaving its own meander loop called as Ox Bow Lake. The valley or flood plain becomes very broad as compared to its depth. The sediment deposit includes gravels in channel section, sands as point bars, silt and clay on flood plain.

As river approaches its mouth and ends at sea or lake, it forms delta through number of distributaries. Such Deltaic River is in its Old Stage where its erosive power is at its minimum, deposition of sediments take place in and around delta while remaining sediment goes into sea. The flood plain becomes extremely wide and is inundated time and again by floods either due to rain in its upper reaches or due to sea tides.

The bridges made on rivers with different stages of flows will create hindrance in natural geomorphic processes operating and will not only restrict river flow but also the dynamics of erosion and deposition. The bridges over rivers of nascent or young stage may have lesser length but will have great heights and may be subjected to flash floods and overtopping. The foundation in all probability will lie on rocks. The nature and thickness of valley fill will show lot of heterogeneity. The bridges on mature stage rivers may have moderate length and height. The problem of flash flood and over topping gets less severe but the current velocity may become very high in rainy seasons as water coming from restricted valleys may spill laterally in these open channels. Hence, the embankments have to be strong and there should be enough space through the bridges, in terms of height of the deck and spacing between the piers to drain off the floodwaters.

The foundation rock may or may not be exposed but can be found at varying depths and can be approached through pile foundations. The bridges on mature or old staged rivers may have great lengths due to broad flood plains but may have comparatively lesser heights. The foundation in this case may have to be provided on river deposits itself as rocks may be found at very deep levels. It is common to have well foundations in such cases. The problem of flash floods and overtopping is minimized.

Problem of Impoundment and Scouring

The height of a bridge is a very important element as for as proper functioning and life of bridge is concerned. During the construction of bridges, roads have to be elevated to the levels of bridge from fairly a long distance and act as barrier to flow of water during flood stage. This results into impoundment of water as in reservoirs causing increased floodwater levels as compared to levels, prior to the construction of bridge at that very place and probable overtopping, which may cause increased buoyancy forces. Similarly, the piers supporting the bridge deck occupy the space of river channel, which in turn will result into raising of water level during lean periods also. During the flood the raised embankment for road and railway to achieve deck height, abutment and piers will also cause damming of flood plain and act as impediment, causing rise in flood water levels as compared to pre bridge construction flood levels.

Hence, before deciding the height of the bridge maximum flood water level is to be worked out by taking into account the probable increase in flood water levels due to construction of road embankment and bridge. Apart from above problem of rise in water level the movement of waters are impeded in channel section, by abutments and piers resulting in to the generation of eddies which causes erosion at the base of abutments and around piers called as scouring, resulting in weakening of their foundation.

Geological Factors

The geological factors play important role where abutment and foundation ground is made up of rocks which is normally a case in hilly areas and rocky areas. The important design factors of a bridge includes total Vertical Load acting downward (Dead Load and Live Load or Load of Rolling Stock i.e. train), Vertical Load acting upward (Buoyancy and Pore Water Pressure), Lateral Forces (Wind Forces, Thrust Forces by running and impounded water) and Dynamic Forces (Vibrations due to movement of four wheelers and trains, Creeping Slopes, Landslides, Earthquake). To address these issues important geological factors to be investigated include type and strength of foundation rocks, thickness and attitude of rocks, structural characteristics of rock mass, groundwater levels, slope strength and earthquake for deciding the site, size, type and design of bridge.

Type and Strength of Foundation Rocks

Depending upon the geology of the area and size of the bridge, foundation rocks at abutment and below piers may remain same or may differ. The former situation is ideal as compared to the later one. Spatial similarity of rocks all through foundation will ensure similar strength, rock mass characteristics and will limit differential settlement. It is the compressive and shear strength of rocks, which are important for deciding types of foundation. Usually all kind of igneous rocks being bodies of large dimension will cover entire foundation area and will have high compressive and shear strength. Only in the case of volcanic rocks inter bedded with weak volcanic ash deposits, care should be taken. Non-foliated metamorphic rocks are as good as igneous rocks as for as their strength and areal coverage is concerned but foliated metamorphic rocks have to be properly investigated, as foliation will impart weakness. If vertical load is incident perpendicular to foliation planes then rock may show comparatively good compressive strength but still shear strength will be low and care should be taken while designing.

Sedimentary rocks are in general show lower strength as compared to the other two major groups primarily due to inherent weakness as well as presence of bedding and lamination planes. Only sandstone with silica and ferruginous cements can be used as foundation ground safely, rest of the rocks such as conglomerate, siltstone, shale etc. may not fair good as foundation rocks especially for medium to large sized bridges. Lime-stones and other calcareous rocks despite being very strong in dry conditions get deteriorated when in contact with water. Water can dissolve calcareous rocks leading to the formation of cavities. Hence, limestone terrains should always be seen with extra caution and construction of bridge may be undertaken only after thorough investigations involving geophysical and drilling methods. The karstic limestone foundation rocks should be treated for its cavities by filling of cement-concrete for safety of a bridge.

Thickness and Attitude of Rocks

Greater is the thickness of rocks, rock foundation will much sounder, as pressure bulb formed due to loading will remain within a particular rock. The natural disposition of rocks may be horizontal, vertical and inclined. The thick horizontal rocks and vertical rocks with different strike directions may fair well as for as foundation of bridges are concerned. The attitude of rocks with respect to bridge alignment becomes important if rocks are inclined wherein the strike direction of rocks will be parallel, perpendicular or diagonal to the bridge alignment. The rocks striking parallel to bridge will have rock dips towards up stream or downstream, the former is a best, while later one is a poor situation. Rocks with high angle dips (>45°) fair good as compared to rocks with low angle of dip (<45°). In the case of rocks striking perpendicular to bridge the rocks may dip either towards left bank or right bank, both the conditions are fair. The rocks with strike oriented diagonally to bridge alignment will also fair good.

Structural Characteristics of Rock Mass

Depending upon the geological evolution of an area, rocks must have undergone natural deformation resulting into development of fractures, shear zone, fault and fold. The number of joints present, their spacing, aperture, alteration etc. will decide the overall strength of rock mass and its engineering behavior. Presence of shear zones in abutments and below piers should be treated by excavation and back filling by reinforced concrete. It has been seen more than often that in hilly areas rivers follow anticlinal fold axis or fault plane/zone because it is easier to erode rocks along them. Depending upon the wavelength and type of fold and overall length of bridge the rocks may get repeated with changing dips resulting into different rocks below different piers and abutments. Similarly, presence of fault along or across the river will result into dislocation of rocks and bringing different rocks at different foundation points apart from forming a deep valley with lot of debris and buried channel. In the geologically active areas, the faults may still show some movement and displacement along them, termed, as Active Faults should be identified as the movement along them may cause failure of the bridge founded over faulted rock mass.

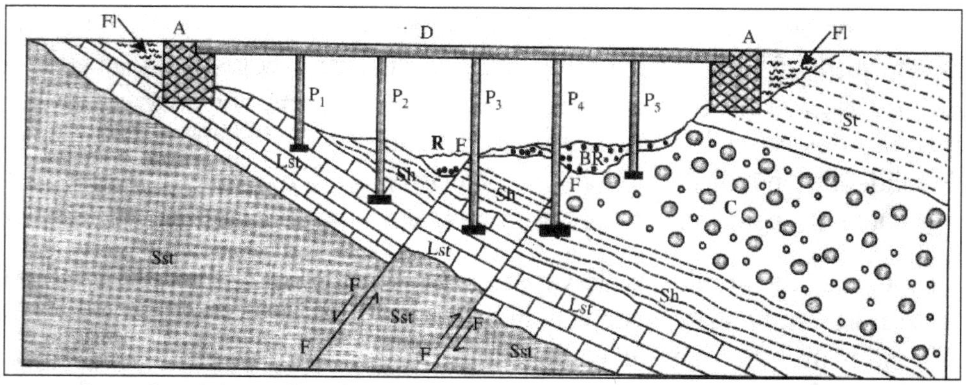

A bridge crossing a river (R) in valley showing faults (F-F) with buried channel (BR) and different rocks. The Abutments (A) and piers (P – P) are placed in different rocks (Sst-sandstone, Lst-limestone, Sh-shale, Cconglomerate, St-siltstone, Fl-earth fill, D-deck).

Groundwater Levels

It is very normal to have groundwater in foundation part of bridge made over a river. The presence of groundwater will result into lowering of shear strength of rocks, increased pore pressure, solution and corrosion of foundation material. If clays are present, in rocks (sedimentary rocks) or as alteration product present in joints they may cause swelling in wet conditions and shrinking during dry seasons may lead to differential settling. Similarly, salts may be found in rocks as minor component in association with calcareous rocks; its dissolution will play havoc with time and may result in the failure of bridge. That is why it will not at all be prudent to have bridge foundation over evaporates, a rock full of salts. Some times during the laying of foundation dewatering is done to lower down the groundwater table well below the foundation and moisture still left is drawn out by injecting hot steam through holes or conducting electro-osmosis by installing electrodes into soil or rocks wherein perforated pipes are as anodes and cathodes to collect water.

Slope Strength

A bridge in hilly areas over a river will definitely be subjected to problems of mass wasting and mass movement may it be very slow termed as creep or fast termed as landslides. In both the cases different elements of bridge will be in danger. There are many instances when landslide in a river valley results into natural damming of river and formation of natural reservoir. The breach of such reservoirs result into flash floods and many times bridge washouts a common feature in Himachal Pradesh and Jammu and Kashmir. To mitigate this problem we have to employ different methods to strengthen the slope as per the need and economy.

Earthquake

For bridge design first of all one has to take into account the seismic zone in which lies the site. The Peak Ground Acceleration (PGA) and probability of earthquakes of different magnitude based on the history of region needs to be considered. The possible return periods of earthquake for 50, 100, 200 years should also be worked out as long-term measures. Apart from this, seismic dampers such as synthetic sheets or ball bearings are placed between the deck and piers to provide base isolation.

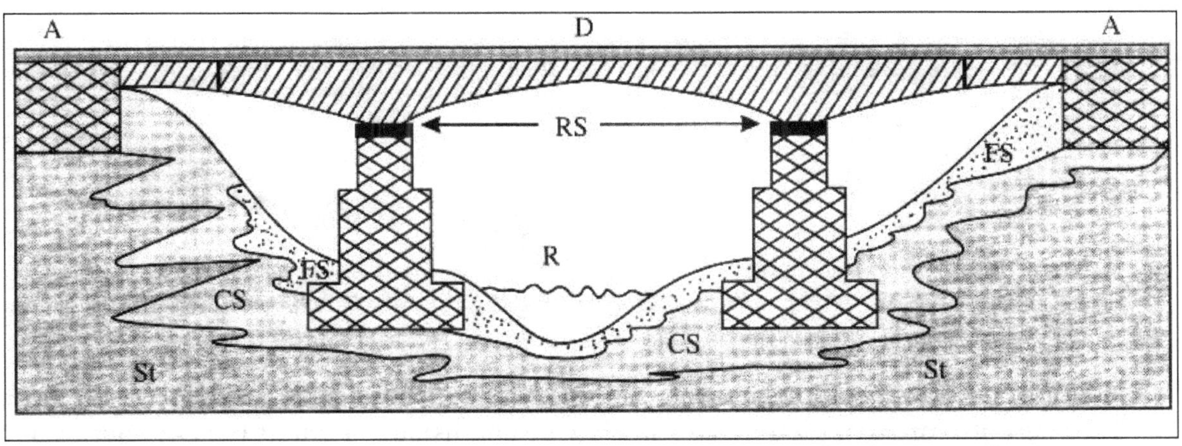

A bridge showing synthetic sheet (RS) between deck and pier (P) to absorb seismic vibrations.

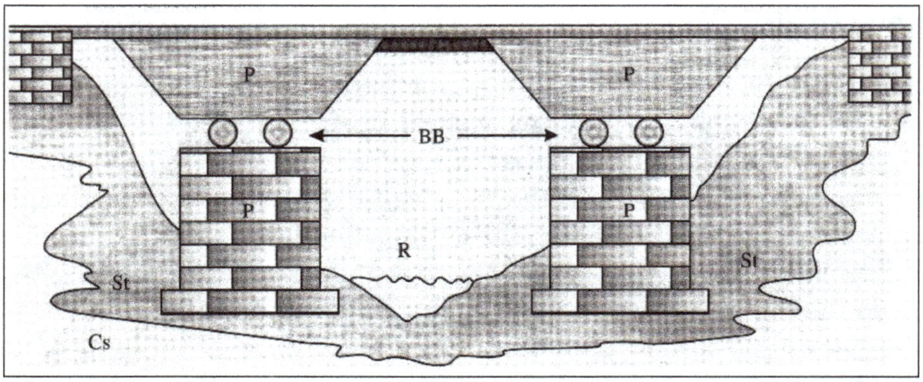

A bridge showing ball bearing (BB) between pre stressed bridge blocks and pier to absorb seismic vibrations.

Earthquake Resistant Structure

Earthquake-resistant structure is Building designed to prevent total collapse, preserve life, and minimize damage in case of an earthquake or tremor. Earthquakes exert lateral as well as vertical forces, and a structure's response to their random, often sudden motions is a complex task that is just beginning to be understood. Earthquake-resistant structures absorb and dissipate seismically induced motion through a combination of means: damping decreases the amplitude of oscillations of a vibrating structure, while ductile materials (e.g., steel) can withstand considerable inelastic deformation. If a skyscraper has too flexible a structure, then tremendous swaying in its upper floors can develop during an earthquake. Care must be taken to provide built-in tolerance for some structural damage, resist lateral loading through stiffeners (diagonal sway bracing), and allow areas of the building to move somewhat independently.

Causes of Earthquakes

Earthquakes are caused by a sudden release of stress along faults in the earth's crust. The continuous motion of tectonic plates causes a steady build-up of pressure in the rock strata on both sides of a fault until the stress is sufficiently great that it is released in a sudden, jerky movement. The resulting waves of seismic energy propagate through the ground and over its surface, causing the shaking we perceive as earthquakes.

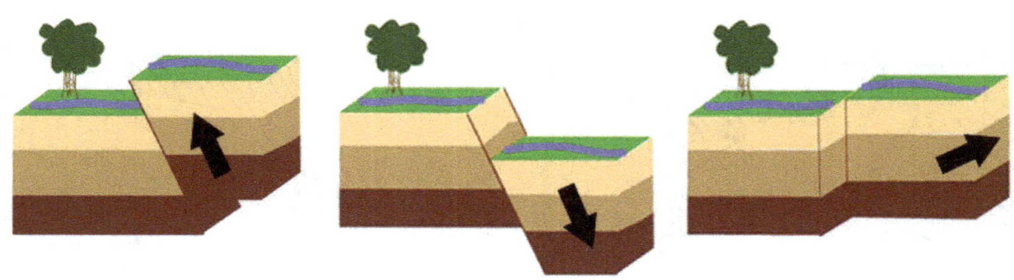
Thrust Fault (convergent) Normal Fault (divergent) Strike-Slip Fault (trasform)

- Tectonic Earthquakes: Earthquakes caused by plate tectonics are called tectonic quakes. They account for most earthquakes worldwide and usually occur at the boundaries of tectonic plates.

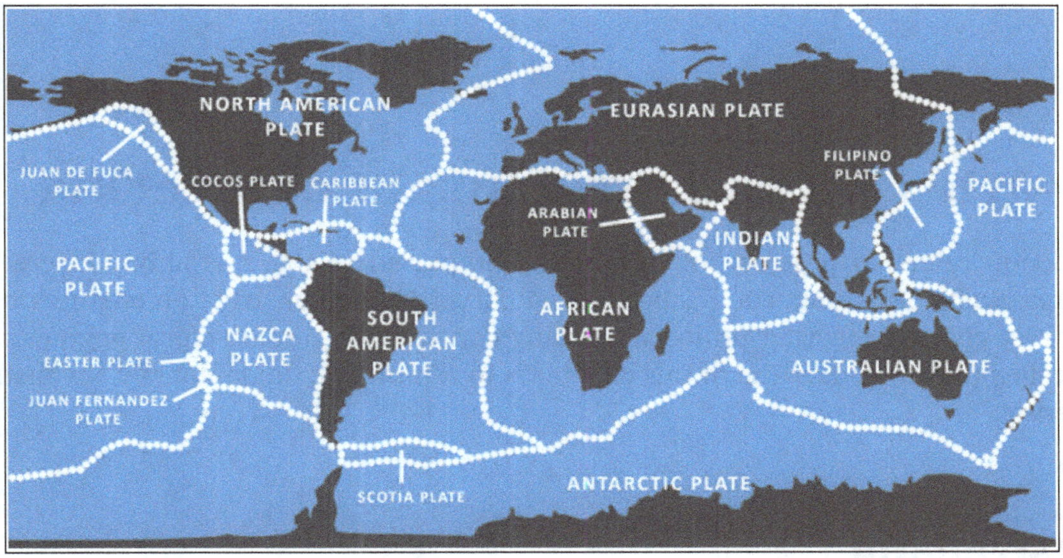

- Induced Earthquakes: Induced quakes are caused by human activity, like tunnel construction, filling reservoirs and implementing geothermal or fracking projects.
- Volcanic Earthquakes: Volcanic quakes are associated with active volcanism. They are generally not as powerful as tectonic quakes and often occur relatively near the surface. Consequently, they are usually only felt in the vicinity of the hypocentre.
- Collapse Earthquakes: Collapse quakes can be triggered by such phenomena as cave-ins, mostly in karst areas or close to mining facilities, as a result of subsidence.

References

- Earthquake-resistant-structure, technology: britannica.com, Retrieved 16, Feb 2020
- Causes-of-earthquakes: seismo.ethz.ch, Retrieved 27, August 2020
- Components-of-dams-and-their-functions: aboutcivil.org, Retrieved 17, June 2020
- Forces-acting-on-a-dam: structuralguide.com, Retrieved 19, July 2020
- Geotechnical_investigations_for_tunnelling-314503138: researchgate.net, Retrieved 07, March 2020
- Technology-terms-and-concepts, science-and-technology: encyclopedia.com, Retrieved 17, May 2020

3

Construction and Structural Engineering

Construction refers to process and techniques employed in the assembly and erection of structures. Some of its important aspects of construction are construction engineering, classification of civil engineering projects, building construction, architectural engineering and structural engineering. The aim of this chapter is to explore the fundamentals of construction engineering.

Construction

In large construction projects, such as this skyscraper in Melbourne, Australia, cranes are essential.

Construction is the process of constructing a building or infrastructure. Construction differs from manufacturing in that manufacturing typically involves [[¥mass production]] of similar items without a designated purchaser, while construction typically takes place on location for a known client. Construction as an industry comprises six to nine percent of the gross domestic product of developed countries. Construction starts with planning, design, and financing; and continues until the project is built and ready for use.

Large-scale construction requires collaboration across multiple disciplines. An architect normally manages the job, and a construction manager, design engineer, construction engineer or project manager supervises it. For the successful execution of a project, effective planning is essential. Those involved with the design and execution of the infrastructure in question must consider zoning requirements, the environmental impact of the job, the successful scheduling, budgeting, construction-site safety, availability and transportation of building materials, logistics, inconvenience to the public caused by construction delays and bidding, etc. The largest construction projects are referred to as megaprojects.

Types of Construction

Industrial assemblage of a thermal oxidizer in the United States of America.

Military residential unit construction by U.S. Navy personnel in Afghanistan.

In general, there are three sectors of construction: buildings, infrastructure and industrial. Building construction is usually further divided into residential and non-residential (commercial/institutional). Infrastructure is often called heavy/highway, heavy civil or heavy engineering. It includes large public works, dams, bridges, highways, water/wastewater and utility distribution. Industrial includes refineries, process chemical, power generation, mills and manufacturing plants. There are other ways to break the industry into sectors or markets.

Engineering News-Record (ENR) is a trade magazine for the construction industry. Each year, ENR compiles and reports on data about the size of design and construction companies. They publish a list of the largest companies in the United States (Top-40) and also a list the largest global firms (Top-250, by amount of work they are doing outside their home country). In 2014, ENR compiled the data in nine market segments. It was divided as transportation, petroleum, buildings, power, industrial, water, manufacturing, sewer/waste, telecom, hazardous waste plus a tenth category for other projects. In their reporting on the Top 400, they used data on transportation, sewer, hazardous waste and water to rank firms as heavy contractors.

The Standard Industrial Classification and the newer North American Industry Classification System have a classification system for companies that perform or otherwise engage in construction. To recognize the differences of companies in this sector, it is divided into three subsectors: building construction, heavy and civil engineering construction, and specialty trade contractors. There are also categories for construction service firms (e.g., engineering, architecture) and construction managers (firms engaged in managing construction projects without assuming direct financial responsibility for completion of the construction project).

Building Construction

Building construction is the process of adding structure to real property or construction of buildings. The majority of building construction jobs are small renovations, such as addition of a room, or renovation of a bathroom. Often, the owner of the property acts as laborer, paymaster, and design team for the entire project. Although building construction projects typically include various common elements, such as design, financial, estimating and legal considerations, many projects of varying sizes reach undesirable end results, such as structural collapse, cost overruns, and/or litigation. For this reason, those with experience in the field make detailed plans and maintain careful oversight during the project to ensure a positive outcome.

The National Cement Share Company of Ethiopia's new plant in Dire Dawa.

Commercial building construction is procured privately or publicly utilizing various delivery methodologies, including cost estimating, hard bid, negotiated price, traditional, management contracting, construction management-at-risk, design & build and design-build bridging.

Residential construction practices, technologies, and resources must conform to local building authority regulations and codes of practice. Materials readily available in the area generally dictate the construction materials used (e.g. brick versus stone, versus timber). Cost of construction on a per square meter (or per square foot) basis for houses can vary dramatically based on site conditions, local regulations, economies of scale (custom designed homes are often more expensive to build) and the availability of skilled tradespeople. As residential construction (as well as all other types of construction) can generate a lot of waste, careful planning again is needed here.

Residential Construction

The most popular method of residential construction in North America is wood-framed construction. Typical construction steps for a single-family or small multi-family house are:

- Develop floor plans and obtain a materials list for estimations (more recently performed with estimating software).
- Obtain government building approval if necessary.
- Clear the building site.

- Survey to stake out for the foundation
- Excavate the foundation and dig footers.
- Pour a foundation and footers with concrete
- Build the main load-bearing structure out of thick pieces of wood and possibly metal I-beams for large spans with few supports.
- Add floor and ceiling joists and install subfloor panels.
- Cover outer walls and roof in OSB or plywood and a water-resistive barrier.
- Install roof shingles or other covering for flat roof.
- Cover the walls with siding, typically vinyl, wood, or brick veneer but possibly stone or other materials.
- Install windows.

Framing.

- Frame interior walls with wooden 2x4s.
- Add internal plumbing, HVAC, electrical, and natural gas utilities.
- Building inspector visits if necessary to approve utilities and framing.
- Install insulation and interior drywall panels (cementboard for wet areas) and to complete walls and ceilings.
- Install bathroom fixtures.
- Spackle, prime, and paint interior walls and ceilings.

- Additional tiling on top of cementboard for wet areas, such as the bathroom and kitchen backsplash.
- Install final floor covering, such as floor tile, carpet, or wood flooring.
- Install major appliances.
- Unless the original owners are building the house, at this point it is typically sold or rented.

New Construction Techniques and Sustainability

As efficiency codes have come into effect in recent years, new construction technologies and methods have emerged. University Construction Management departments are on the cutting edge of the newest methods of construction intended to improve efficiency, performance and reduce construction waste.

New techniques of building construction are being researched, made possible by advances in 3D printing technology. In a form of *additive building construction*, similar to the additive manufacturing techniques for manufactured parts, building printing is making it possible to flexibly construct small commercial buildings and private habitations in around 20 hours, with built-in plumbing and electrical facilities, in one continuous build, using large 3D printers. Working versions of 3D-printing building technology are already printing 2 metres (6 ft 7 in) of building material per hour as of January 2013, with the next-generation printers capable of 3.5 metres (11 ft) per hour, sufficient to complete a building in a week. Dutch architect Janjaap Ruijssenaars's performative architecture 3D-printed building was scheduled to be built in 2014.

In the current trend of sustainable construction, the recent movements of New Urbanism and New Classical Architecture promote a sustainable approach towards construction, that appreciates and develops smart growth, architectural tradition and classical design. This is in contrast to modernist and short-lived globally uniform architecture, as well as opposing solitary housing estates and suburban sprawl. Both trends started in the 1980s.

The construction site may be shut down due to bad weather. Erecting scaffolded tents over the site may reduce the number of lost work days, increasing productivity.

Construction Processes

Design Team

In the modern industrialized world, construction usually involves the translation of designs into reality. A formal design team may be assembled to plan the physical proceedings, and to integrate those proceedings with the other parts. The design usually consists of drawings and specifications, usually prepared by a design team including Architect, civil engineers, mechanical engineers, electrical engineers, structural engineers, fire protection engineers, planning consultants, architectural consultants, and archaeological consultants. The design team is most commonly employed by (i.e. in contract with) the property owner. Under this system, once the design is completed by the design team, a number of construction companies or construction management companies may then be asked to make a bid for the work, either based directly on the design, or on the basis of drawings and a bill of quantities provided by a quantity surveyor. Following evaluation of bids, the owner typically awards a contract to the most cost efficient bidder.

Shasta Dam under construction in June 1942.

The best modern trend in design is toward integration of previously separated specialties, especially among large firms. In the past, architects, interior designers, engineers, developers, construction managers, and general contractors were more likely to be entirely separate companies, even in the larger firms. Presently, a firm that is nominally an "architecture" or "construction management" firm may have experts from all related fields as employees, or to have an associated company that provides each necessary skill. Thus, each such firm may offer itself as "one-stop shopping" for a construction project, from beginning to end. This is designated as a "design build" contract where the contractor is given a performance specification and must undertake the project from design to construction, while adhering to the performance specifications.

Several project structures can assist the owner in this integration, including design-build, partnering and construction management. In general, each of these project structures allows the owner to integrate the services of architects, interior designers, engineers and constructors throughout design and construction. In response, many companies are growing beyond traditional offerings of design or construction services alone and are placing more emphasis on establishing relationships with other necessary participants through the design-build process.

The increasing complexity of construction projects creates the need for design professionals trained in all phases of the project's life-cycle and develop an appreciation of the building as an advanced technological system requiring close integration of many sub-systems and their individual components, including sustainability. Building engineering is an emerging discipline that attempts to meet this new challenge.

Financial Advisors

Trump International Hotel and Tower (Chicago)

Construction projects can suffer from preventable financial problems. Underbids happen when builders ask for too little money to complete the project. Cash flow problems exist when the present amount of funding cannot cover the current costs for labour and materials, and because they are a matter of having sufficient funds at a specific time, can arise even when the overall total is enough. Fraud is a problem in many fields, but is notoriously prevalent in the construction field. Financial planning for the project is intended to ensure that a solid plan with adequate safeguards and contingency plans are in place before the project is started and is required to ensure that the plan is properly executed over the life of the project.

May 23, 2006.

September 14, 2007 (3 months before completion).

Mortgage bankers, accountants, and cost engineers are likely participants in creating an overall plan for the financial management of the building construction project. The presence of the mortgage banker is highly likely, even in relatively small projects since the owner's equity in the property is the most obvious source of funding for a building project. Accountants act to study the expected monetary flow over the life of the project and to monitor the payouts throughout the process. Cost engineers and estimators apply expertise to relate the work and materials involved to a proper valuation. Cost overruns with government projects have occurred when the contractor identified change orders or project changes that increased costs, which are not subject to competition from other firms as they have already been eliminated from consideration after the initial bid.

Large projects can involve highly complex financial plans and often start with a conceptual estimate performed by a building estimator. As portions of a project are completed, they may be sold, supplanting one lender or owner for another, while the logistical requirements of having the right trades and materials available for each stage of the building construction project carries forward. In many English-speaking countries, but not the United States, projects typically use quantity surveyors.

Legal Aspects

Construction along Ontario Highway 401, widening the road from six to twelve travel lanes.

Construction site with tubes.

A construction project must fit into the legal framework governing the property. These include governmental regulations on the use of property, and obligations that are created in the process of construction.

The project must adhere to zoning and building code requirements. Constructing a project that fails to adhere to codes does not benefit the owner. Some legal requirements come from malum in se considerations, or the desire to prevent things that are indisputably bad – bridge collapses or explosions. Other legal requirements come from malum prohibitum considerations, or things that are a matter of custom or expectation, such as isolating businesses to a business district and residences to a residential district. An attorney may seek changes or exemptions in the law that governs the land where the building will be built, either by arguing that a rule is inapplicable (the bridge design will not cause a collapse), or that the custom is no longer needed (acceptance of live-work spaces has grown in the community).

A construction project is a complex net of contracts and other legal obligations, each of which all parties must carefully consider. A contract is the exchange of a set of obligations between two or more parties, but it is not so simple a matter as trying to get the other side to agree to as much as possible in exchange for as little as possible. The time element in construction means that a delay costs money, and in cases of bottlenecks, the delay can be extremely expensive. Thus, the contracts must be designed to ensure that each side is capable of performing the obligations set out. Contracts that set out clear expectations and clear paths to accomplishing those expectations are far more likely to result in the project flowing smoothly, whereas poorly drafted contracts lead to confusion and collapse.

Legal advisors in the beginning of a construction project seek to identify ambiguities and other potential sources of trouble in the contract structure, and to present options for preventing problems. Throughout the process of the project, they work to avoid and resolve conflicts that arise. In each case, the lawyer facilitates an exchange of obligations that matches the reality of the project.

Interaction of Expertise

Apartment complex under construction in Daegu, South Korea.

Design, finance, and legal aspects overlap and interrelate. The design must be not only structurally sound and appropriate for the use and location, but must also be financially possible to build, and legal to use. The financial structure must accommodate the need for building the design provided, and must pay amounts that are legally owed. The legal structure must integrate the design into the surrounding legal framework, and enforce the financial consequences of the construction process.

Procurement

Procurement describes the merging of activities undertaken by the client to obtain a building. There are many different methods of construction procurement; however the three most common types of procurement are traditional (design-bid-build), design-build and management contracting.

There is also a growing number of new forms of procurement that involve relationship contracting where the emphasis is on a co-operative relationship between the principal and contractor and other stakeholders within a construction project. New forms include partnering such as Public-Private Partnering (PPPs) aka private finance initiatives (PFIs) and alliances such as "pure" or "project" alliances and "impure" or "strategic" alliances. The focus on co-operation is to ameliorate the many problems that arise from the often highly competitive and adversarial practices within the construction industry.

Traditional

This is the most common method of construction procurement and is well established and recognized. In this arrangement, the architect or engineer acts as the project coordinator. His or her role is to design the works, prepare the specifications and produce construction drawings, administer the contract, tender the works, and manage the works from inception to completion. There are direct contractual links between the architect's client and the main contractor. Any subcontractor has a direct contractual relationship with the main contractor. The procedure continues until the building is ready to occupy.

Design-build

Construction of the *Phase-1* (first two towers) of the Havelock City Project, Sri Lanka.

This approach has become more common in recent years, and involves the client contracting a single entity to both provide a design and to build that design. In some cases, the design-build package can also include finding the site, arranging funding and applying for all necessary statutory consents.

The owner produces a list of requirements for a project, giving an overall view of the project's goals. Several D&B contractors present different ideas about how to accomplish these goals. The owner selects the ideas he or she likes best and hires the appropriate contractor. Often, it is not just one contractor, but a consortium of several contractors working together. Once these have been hired, they begin building the first phase of the project. As they build phase 1, they design phase 2. This is in contrast to a design-bid-build contract, where the project is completely designed by the owner, then bid on, then completed.

Kent Hansen pointed out that state departments of transportation usually use design build contracts as a way of progressing projects when states lack the skills-resources. In such departments, design build contracts are usually employed for very large projects.

Management Procurement Systems

In this arrangement the client plays an active role in the procurement system by entering into separate contracts with the designer (architect or engineer), the construction manager, and individual trade contractors. The client takes on the contractual role, while the construction or project manager provides the active role of managing the separate trade contracts, and ensuring that they complete all work smoothly and effectively together.

Management procurement systems are often used to speed up the procurement processes, allow the client greater flexibility in design variation throughout the contract, give the ability to appoint individual work contractors, separate contractual responsibility on each individual throughout the contract, and to provide greater client control.

In recent time, construction software starts to get traction - as it digitizes construction industry. Among solutions, there are for example: Procore, GenieBelt, PlanGrid, etc.

Authority Having Jurisdiction

In construction, the authority having jurisdiction (AHJ) is the governmental agency or sub-agency that regulates the construction process. In most cases, this is the municipality where the building is located. However, construction performed for supra-municipal authorities are usually regulated directly by the owning authority, which becomes the AHJ.

Construction on the Federal Reserve building in Kansas City, Missouri.

Before the foundation can be dug, contractors are typically required to verify and have existing utility lines marked, either by the utilities themselves or through a company specializing in such services. This lessens the likelihood of damage to the existing electrical, water, sewage, phone, and cable facilities, which could cause outages and potentially hazardous situations. During the construction of a building, the municipal building inspector inspects the building periodically to ensure that the construction adheres to the approved plans and the local building code. Once construction is complete and a final inspection has been passed, an occupancy permit may be issued.

An operating building must remain in compliance with the fire code. The fire code is enforced by the local fire department or a municipal code enforcement office.

Changes made to a building that affect safety, including its use, expansion, structural integrity, and fire protection items, usually require approval of the AHJ for review concerning the building code.

Industry Characteristics

In the United States, the industry in 2014 has around $960 billion in annual revenue according to statistics tracked by the Census Bureau, of which $680 billion is private (split evenly between residential and nonresidential) and the remainder is government. As of 2005, there were about 667,000 firms employing 1 million contractors (200,000 general contractors, 38,000 heavy, and 432,000 specialty); the average contractor employed fewer than 10 employees. As a whole, the industry employed an estimated 5.8 million as of April 2013, with a 13.2% unemployment rate. In the United States, approximately 828,000 women were employed in the construction industry as of 2011.

Careers

Helicopter view of the Atacama Large Millimeter/submillimeter Array (ALMA) Operations Support Facility (OSF) construction site.

Ironworkers erecting the steel frame of a new building at Massachusetts General Hospital in Boston.

Skilled occupations include carpenters, electricians, plumbers, ironworkers, masons, and many other manual crafts, as well as those involved in project management. In the UK these require further education qualifications, often in vocational subject areas. These qualifications are either obtained directly after the completion of compulsory education or through "on the job" apprenticeship training. In the UK, 8500 construction-related apprenticeships were commenced in 2007.

A truck operator at Al Gamil, the largest construction company in Djibouti.

There are many routes to the different careers within the construction industry. These three main tiers are based on educational background and training, which vary by country:

- Unskilled and semi-skilled – General site labor with little or no construction qualifications.
- Skilled – Tradesmen who've served apprenticeships, typically in labor unions, and on-site managers who possess extensive knowledge and experience in their craft or profession.
- Technical and management – Personnel with the greatest educational qualifications, usually graduate degrees, trained to design, manage and instruct the construction process.

Technical and specialized occupations require more training as a greater technical knowledge is required. These professions also hold more legal responsibility. A short list of the main careers with an outline of the educational requirements are given below:

- Architect – Typically holds 1, undergraduate 3 year degree in architecture + 1, post-graduate 2 year degree (DipArch or BArch) in architecture plus 24 months experience within the industry. To use the title «architect» the individual must be registered on the Architects Registration Board register of Architects.
- Civil engineer – Typically holds a degree in a related subject. The Chartered Engineer qualification is controlled by the Engineering Council, and is often achieved through membership of the Institution of Civil Engineers. A new university graduate must hold a master's degree to become chartered; persons with bachelor's degrees may become an Incorporated Engineer.
- Building services engineer – Often referred to as an «M&E Engineer» typically holds a degree in mechanical or electrical engineering. Chartered Engineer status is governed by the Engineering Council, mainly through the Chartered Institution of Building Services Engineers.
- Project manager – Typically holds a 4-year or greater higher education qualification, but are often also qualified in another field such as architecture, civil engineering or quantity surveying.

- Structural engineer – Typically holds a bachelor›s or master›s degree in structural engineering. A P.ENG is required from the Professional Engineers Ontario (Canada). New university graduates must hold a master›s degree to gain chartered status from the Engineering Council, mainly through the Institution of Structural Engineers (UK).
- Quantity surveyor – Typically holds a bachelor›s degree in quantity surveying. Chartered status is gained from the Royal Institution of Chartered Surveyors.
- Civil estimators are professionals who typically have a background in civil engineering, construction project management, or construction supervision.

In 2010 a salary survey revealed the differences in remuneration between different roles, sectors and locations in the construction and built environment industry. The results showed that areas of particularly strong growth in the construction industry, such as the Middle East, yield higher average salaries than in the UK for example. The average earning for a professional in the construction industry in the Middle East, across all sectors, job types and levels of experience, is £42,090, compared to £26,719 in the UK. This trend is not necessarily due to the fact that more affluent roles are available, however, as architects with 14 or more years experience working in the Middle East earn on average £43,389 per annum, compared to £40,000 in the UK. Some construction workers in the US/Canada have made more than $100,000 annually, depending on their trade.

Safety

At-risk workers without appropriate safety equipment.

Construction is one of the most dangerous occupations in the world, incurring more occupational fatalities than any other sector in both the United States and in the European Union. In 2009, the fatal occupational injury rate among construction workers in the United States was nearly three times that for all workers. Falls are one of the most common causes of fatal and non-fatal injuries among construction workers. Proper safety equipment such as harnesses and guardrails and procedures such as securing ladders and inspecting scaffolding can curtail the risk of occupational injuries in the construction industry. Other major causes of fatalities in the construction industry include electrocution, transportation accidents, and trench cave-ins.

Roofing requires a very high level of safety.

Other safety risks for workers in construction include hearing loss due to high noise exposure, musculoskeletal injury, chemical exposure, and high levels of stress.

History

A picture of a building under construction in India.

The first huts and shelters were constructed by hand or with simple tools. As cities grew during the Bronze Age, a class of professional craftsmen, like bricklayers and carpenters, appeared. Occasionally, slaves were used for construction work. In the Middle Ages, these were organized into guilds. In the 19th century, steam-powered machinery appeared, and later diesel- and electric powered vehicles such as cranes, excavators and bulldozers.

Fast-track construction has been increasingly popular in the 21st century. Some estimates suggest that 40% of construction projects are now fast-track construction.

Construction Phases

- Vision/fantasy/idea - a concept never intended to be built, may be an aesthetic or structural design exercise.
- Proposed - a building concept that is under review by the building owner and by government.
- Approved - a building concept that will be constructed in the near future. If the proposed building is not approved then the proposal may be amended and resubmitted, or it may be deferred or cancelled.
- Design - the specification of what is to be built in sufficient detail to be used as the basis as a contract between the owner and a contractor.
- Procurement - the selection of the contractor or contractors to carry out the construction. This may be by competitive tendering.
- Diversions - before construction can start any services on the site which must be kept operational to serve other adjacent sites must be diverted so they run outside the footprint of the new building. This can include drainage, water and gas piped services and power and communication cables.
- Under-construction - a fully designed building currently being built.
- Ground works - construction work below ground level including the construction of basements and foundations.
- Topped-out - a fully designed building where construction has reached the highest point of the building.
- Fitting out - installation of the decorative, non-structural elements once the building main structure is complete. This includes painting, ceilings, light fittings etc.
- Commissioning or setting to work - Once the building Mechanical, electrical, plumbing, communications, and building control systems are installed they then need to be tested and adjusted so they deliver the required performance. In modern buildings this can take some time during which little seems to be going on but if this is not done properly then these systems will not deliver their design performance leading to hot and cold spots, spurious alarms, higher energy bills, and systems failing during emergencies.
- Substantial Completion / Beneficial Occupancy - a point when the work is sufficiently complete so that the Owner can occupy (Items noted during inspection 'punch list' or 'snag list' may still be corrected).
- Complete/built - a fully designed building that has been fully built, excluding future expansions (punch list items all completed).
- Building Operation - All those day-to-day activities need to ensure the building can be used. In simple buildings this means little more than cleaning but in more complicated buildings this is a large scale operation employing a large team of staff. If they do their job right then you hardly notice them.

- Maintenance - works to ensure the building continues to operate in accordance with its design, including replacing elements which are approaching the end of their useful life.
- Repair - replacing building elements which have been damaged or which have failed to restore the building to its as-built state.
- Renovation - modification to the building. This can be minor modifications that are carried out while the building is occupied or major works where only the structural elements are kept and the building is out of use for years.
- Demolition - destruction of the building which may include the salvage of some elements for reuse elsewhere.

Engineering for Construction

Construction engineering is a professional discipline that deals with the designing, planning, construction, and management of infrastructures such as roads, tunnels, bridges, airports, railroads, facilities, buildings, dams, utilities and other projects. Construction engineering is considered a professional sub-practice area of Civil engineering.

Construction technology is a related field that deals more with the practical aspects of projects. Construction technologists or construction technicians learn some of the design aspects similar to civil engineers and some of the project site management aspects similar to construction managers. These technicians are unique such that they are a cross between civil engineers and construction managers.

At the educational level, civil engineering students concentrate primarily on the design work, which is more analytical, gearing them toward a career as a design professional. This essentially requires them to take a multitude of challenging engineering science and design courses as part of obtaining a 4 year accredited degree. Also, education for construction managers is primarily focused on construction procedures, methods, costs, schedules, and personnel management. Their primary concern is to deliver a project on time, within budget, and of the desired quality.

The difference between a construction technologist and a civil engineer is that civil engineering is an engineering discipline. Construction technology students take basic design courses as well as construction management courses. This allows them to understand both basic design functions as well as the building requirements needed to design and build today's infrastructures.

Work Activities

Depending on which career the construction engineer has chosen to follow, an entry-level design engineer normally provides support to project managers and assist with creating conceptual designs, scopes, and cost estimates for the planning and construction of approved projects. It should be noted that a career in design work does require a professional engineer license (PE). Individuals who pursue this career path are strongly advised to sit for the Engineer In Training exam (EIT) while in college as it takes five years (4 years in USA) post graduate to obtain the PE license.

Entry-level construction manager positions are typically called project engineers or assistant project engineers. They are responsible for preparing purchasing requisitions, processing change orders, preparing monthly budgeting reports, and handling meeting minutes. The construction management position does not necessarily require a PE license; however possessing one does make the individual more marketable, as the PE license allows the individual to sign off on temporary structure designs.

Abilities

Construction engineers are problem solvers. They contribute to the creation of infrastructure that best meets the unique demands of its environment. They must be able to understand infrastructure life cycles. When compared and contrasted to design engineers, construction engineers bring to the table their own unique perspectives for solving technical challenges with clarity and imagination. While individuals considering this career path should certainly have a strong understanding of mathematics and science, many other skills are also highly desirable, including critical and analytical thinking, time management, people management, and good communication skills.

Educational Requirements

Individuals looking to obtain a construction engineering degree must first ensure that the program is accredited by EAC or Technology Accreditation Commission (TAC) of the Accreditation Board for Engineering and Technology (ABET). ABET accreditation is assurance that a college or university program meets the quality standards established by the profession for which it prepares its students. In the US there are currently twenty-five programs that exist in the entire country so careful college consideration is advised.

A typical construction engineering curriculum is a mixture of engineering mechanics, engineering design, construction management and general science and mathematics. This usually leads to a Bachelor of Science degree. The B.S. degree along with some design or construction experience is sufficient for most entry level positions. Graduate schools may be an option for those who want to go further in depth of the construction and engineering subjects taught at the undergraduate level. In most cases construction engineering graduates look to either civil engineering, engineering management, or business administration as a possible graduate degree.

Job Prospects

Job prospects for construction engineers generally have a strong cyclical variation. For example, starting in 2008 - continuing until at least 2011 - job prospects have been poor due to the collapse of housing bubbles in many parts of the world. This sharply reduced demand for construction, forced construction professionals towards infrastructure construction and therefore increased the competition faced by established and new construction engineers. This increased competition, and a core reduction in quantity demand is in parallel with a possible shift in the demand for construction engineers due to the automation of many engineering tasks, overall resulting in reduced prospects for construction engineers. In early 2010 the United States construction industry had a 27% unemployment rate, this is nearly three times higher than the 9.7% national average unemployment rate. The construction unemployment rate (including tradesmen) is comparable to the United States 1933 unemployment rate - the lowest point of the Great Depression - of 25%.

Remuneration

The average salary for a civil engineer in the UK depends on the sector, and more specifically the level of experience of the individual. A 2010 survey of the remuneration and benefits of those occupying jobs in construction and the built environment industry showed that the average salary of a civil engineer in the UK is £29,582. In the United States, as of May 2013, the average was $85,640. The average salary varies depending on experience, for example the average annual salary for a civil engineer with between 3 and 6 years experience is £23,813. For those with between 14 and 20 years experience the average is £38,214.

The art of construction has slowly developed into a science over the centuries. Civil engineers are involved from conception to physical realization of the project. We shall first enumerate the different civil engineering projects.

Building Project

1. Residential Building, Town Planning

Govt Staff Quaters.	Residential Buildings.		

2. Other Buildings

Academic building such as institutes, university and schools, hospitals, temples, government and private offices, commercial complexes, recreational building such as auditorium, cinema hall, theaters, stores, warehouse, light manufacturing plants.

Guest House.	School.	Hospital.	Temple.

Depending upon the level of development of a country, these construction projects constitute large share. Developed countries have more facilities in this area than porly developed countries.

Transportation Systems

Bridges and flyover, roads, railways, airport, urban mass transit system such as mono railway and ropeways , town planning, port and harbour, tunnel.

Bridges & Flyover.	Airport.	Road.

These are basic infrastructural projects. Developing countries are spending major of its budget on these projects.

Hydraulic Structures

Dam, barrage, canal, hydroelectric power plant, tunnel, embankment, flood control system.

Dam.	Canal.	Barrage.

For a country whose economy is dependent on agriculture, these are major infrastructural projects. India spent almost 1/5 th of its planning budget on these projects in first and second five year plan.

Fossil Fuel and Nuclear Power Plants

Nuclear Plant.

Thermal Power Plants.

Energy is vital for survival. The generation of electricity is carried out from coal, petroleum and nuclear fuel.

Offshore Stuctures

Tension Legged Platform.

These are huge steel and concrete structure in sea, used usually for petroleum extraction.

Space Based and Defense Structure

Army Cantonment, missile launch pad, bunkers.

Army Cantonment.

Missle Launch Pad.

Defence is an important part of any country. Thses are huge civil engineering construction projects related to defence. Technology involved is very sophisticated.

Industrial Structures

Petrochemical refinaries, small and large scale manufacturing plants, silos, bunker, mine developments.

Refinary.		

These are the backbone of modern growth of a country.

Communication System

Transmission line tower, communication network, TV Tower.

Transmission Line.	TV Tower.	

These structures are lifeline for present activities.

Water and Waste Project

- Water treatment plant.
- Water distribution system - pipe, water tanks.
- Sewage and storm water collection system.

- Sewage treatment plants.
- Sewage disposal system.

Water Treatment.	Waste Water Treatment.	Effulent Treatment.

These structures are important for environment friendly living and healthy life.

Special Projects

High altitude construction project, marine exploration project.

High altitude construction.

These are used for scientific exploration, strategic importance.

Classification of Civil Engineering Projects

Civil Engineer Project manager or construction manager has to interact with various field of specializations depending upon the type of the project.

A project manager involved in the construction of hospital has to interact with medical professional.

Similarly, the construction of petroleum, refinery involves the complex interaction with chemical engineer, mechanical engineer, electrical engineers.

Based on the intersection of knowledge and interaction with other disciplines, civil engineering projects can be grouped as follows:

- Building Projects.
- Heavy engineering Project.
- Industrial Projects.

Building construction projects can be further divided into Residential Buildings and commercial buildings. We shall discuss the nature of the project next :

Building Construction

During design phase, building design is coordinated by architect who works with other engineering disciplines such as structure engineer, electrical engineer whereas construction phase is coordinated by the civil engineers. But group leader in the building projects is generally architect. Building project can further be classified into:

- Residential Building.
- Commercial and other buildings.
- Residential buildings in India are primarily single or double storey building. Multi-storied residential buildings are also getting constructed. Other building project produces variety of building ranging from hospital, temples, and small stores to commercial office tower, theatres and cinema hall, government buildings, recreational facilities, small schools to new institute like IIT or university.
- Economically building project accounts for around 60 to 70 percent of construction project in a country. But most of the constructions are carried out in an unorganized way. In India, building construction projects are very fragmented and majority of construction in rural and semi urban area is carried out by non-engineering ways. Building construction in India is highly labour –intensive. Most of the labourers employed for building projects are semi-skilled and untrained. Uses of heavy equipments are very rare. Building projects are governed by many local as well as general laws, such as municipal laws, building laws etc.

Building

A building or edifice is a structure with a roof and walls standing more or less permanently in one place, such as a house or factory. Buildings come in a variety of sizes, shapes and functions, and have been adapted throughout history for a wide number of factors, from building materials available, to weather conditions, to land prices, ground conditions, specific uses and aesthetic reasons. To better understand the term *building* compare the list of nonbuilding structures.

Buildings serve several needs of society – primarily as shelter from weather, security, living space, privacy, to store belongings, and to comfortably live and work. A building as a shelter represents a physical division of the human habitat (a place of comfort and safety) and the *outside* (a place that at times may be harsh and harmful).

The Hôtel de Ville, building in Paris, France.

A Shaolin Monastery building in China.

Ever since the first cave paintings, buildings have also become objects or canvasses of much artistic expression. In recent years, interest in sustainable planning and building practices has also become an intentional part of the design process of many new buildings.

Definitions

A building and skybridge in Munich, Germany.

The Rotunda in Hellerup, Denmark, a cylindric building made in steel frame and aluminum.

Example of a religious building: the Great Mosque of Kairouan (also called the Mosque of Uqba), founded in 670, dates in its present state principally from the 9th century. The Great Mosque of Kairouan is located in the city of Kairouan, Tunisia.

Da Vinci building (at the left), in Mendoza, Argentina. Considered one of the 1000 architectures of the Americas.

The word *building* is both a noun and a verb also an adverb: the structure itself and the act of making it. As a noun, a building is 'a structure that has a roof and walls and stands more or less permanently in one place'; "there was a three-storey building on the corner"; "it was an imposing edifice". In the broadest interpretation a fence or wall is a building. However, the word *structure* is used more broadly than *building* including natural and man-made formations and does not necessarily have walls. Structure is more likely to be used for a fence. Sturgis' Dictionary included that "[building] differs from architecture in excluding all idea of artistic treatment; and it differs from construction in the idea of excluding scientific or highly skilful treatment." As a verb, building is the act of construction.

Structural height in technical usage is the height to the highest architectural detail on building from street-level. Depending on how they are classified, spires and masts may or may not be included in this height. Spires and masts used as antennas are not generally included. The definition of a *low-rise vs. a high-rise* building is a matter of debate, but generally three storeys or less is considered low-rise.

History

A report by Shinichi Fujimura of a shelter built 500 000 years ago is doubtful since Fujimura was later found to have faked many of his findings. Supposed remains of huts found at the Terra Amata site in Nice purportedly dating from 200 000 to 400 000 years ago have also been called into question. There is clear evidence of homebuilding from around 18 000 BC. Buildings became common during the Neolithic.

Types

A timber-framed house in Marburg, Germany.

A block of tenements (apartments) in Bruntsfield, Edinburgh, Scotland.

Residential

Single-family residential buildings are most often called houses or homes. Residential buildings containing more than one dwelling unit are called a duplex, apartment building to differentiate them from 'individual' houses. A condominium is an apartment that the occupant owns rather than rents. Houses may also be built in pairs (semi-detached), in terraces where all but two of the houses have others either side; apartments may be built round courtyards or as rectangular blocks surrounded by a piece of ground of varying sizes. Houses which were built as a single dwelling may later be divided into apartments or bedsitters; they may also be converted to another use e.g. an office or a shop.

Building types may range from huts to multimillion-dollar high-rise apartment blocks able to house thousands of people. Increasing settlement density in buildings (and smaller distances between buildings) is usually a response to high ground prices resulting from many people wanting to live close to work or similar attractors. Other common building materials are brick, concrete or combinations of either of these with stone.

Residential buildings have different names for their use depending if they are seasonal include holiday cottage (vacation home) or timeshare; size such as a cottage or great house; value such as a shack or mansion; manner of construction such as a log home or mobile home; proximity to the ground such as earth sheltered house, stilt house, or tree house. Also if the residents are in need of special care such as a nursing home, orphanage or prison; or in group housing like barracks or dormitories.

Historically many people lived in communal buildings called longhouses, smaller dwellings called pit-houses and houses combined with barns sometimes called housebarns.

Buildings are defined to be substantial, permanent structures so other dwelling forms such as houseboats, yurts, and motorhomes are dwellings but not buildings.

Multi-storey

A multi-storey is a building that has multiple floors.

Complex

Sometimes a group of inter-related (and possibly inter-connected) builds are referred to as a complex - for example a *housing complex, educational complex, hospital complex*, etc.

Creation

The practice of designing, constructing, and operating buildings is most usually a collective effort of different groups of professionals and trades. Depending on the size, complexity, and purpose of a particular building project, the project team may include:

- A real estate developer who secures funding for the project.
- One or more financial institutions or other investors that provide the funding.
- Local planning and code authorities.
- A Surveyor who performs an ALTA/ACSM and construction surveys throughout the project.

- Construction managers who coordinate the effort of different groups of project participants.
- Licensed architects and engineers who provide building design and prepare construction documents.
- The principal design Engineering disciplines which would normally include the following professionals:- Civil, Structural, Mechanical building services or HVAC (heating Ventilation and Air Conditioning) Electrical Building Services, Plumbing and drainage. Also other possible design Engineer specialists may be involved such as Fire (prevention), Acoustic, facade engineers,building physics,Telecomms, AV (Audio Visual), BMS (Building Management Systems)Automatic controls etc. These design Engineers also prepare construction documents which are issued to specialist contractors to obtain a price for the works and to follow for the installations.
- Landscape architects.
- Interior designers.
- Other consultants.
- Contractors who provide construction services and install building systems such as climate control, electrical, plumbing, Decoration, fire protection, security and telecommunications.
- Marketing or leasing agents.
- Facility managers who are responsible for operating the building.

Regardless of their size or intended use, all buildings in the US must comply with zoning ordinances, building codes and other regulations such as fire codes, life safety codes and related standards.

Vehicles—such as trailers, caravans, ships and passenger aircraft—are treated as "buildings" for life safety purposes.

Ownership and Funding

- Mortgage loan.
- Real estate developer.

Building Services

Physical Plant

Any building requires a certain amount of internal infrastructure to function, which includes such elements like heating / cooling, power and telecommunications, water and wastewater etc. Especially in commercial buildings (such as offices or factories), these can be extremely intricate systems taking up large amounts of space (sometimes located in separate areas or double floors / false ceilings) and constitute a big part of the regular maintenance required.

The BB&T Building in Macon, Georgia is constructed of aluminum.

Conveying Systems

Systems for transport of people within buildings:

- Elevator
- Escalator
- Moving sidewalk (horizontal and inclined)

Systems for transport of people between interconnected buildings:

- Skyway
- Underground city

Building Damage

A building in Massueville, Quebec, Canada engulfed by fire.

Buildings may be damaged during the construction of the building or during maintenance. There are several other reasons behind building damage like accidents such as storms, explosions and subsidence caused by mining or poor foundations. Buildings also may suffer from fire damage and flooding in special circumstances. They may also become dilapidated through lack of proper maintenance or alteration work improperly carried out.

Tower Block

A high-rise residential apartment building in Tseung Kwan O, Hong Kong.

A tower block, high-rise, apartment tower, residential tower, apartment block, block of flats, or office tower is a tall building or structure used as a residential and/or office building. In some areas it may be referred to as an "MDU", standing for "Multi Dwelling Unit". In the United States, such a structure is referred to as an apartment building or office building, while a group of such buildings is called an apartment complex or office complex.

One of a pair of 47-story high-rise residential complexes, a ubiquitous feature of the borough of Fort Lee, Bergen County, New Jersey, United States, in the New York City Metropolitan Area, with several over 90 metres (300 ft) tall.

There are numerous high-rise apartments in Navi Mumbai, India.

High-rise buildings became possible with the invention of the elevator (lift) and cheaper, more abundant building materials. The materials used for the structural system of high-rise buildings are reinforced concrete and steel. Most North American style skyscrapers have a steel frame, while residential blocks are usually constructed of concrete. There is no clear difference between a tower block and a skyscraper, although a building with fifty or more stories is generally considered a skyscraper.

10 story high-rise in Manchester, New Hampshire, U.S.

High-rise structures pose particular design challenges for structural and geotechnical engineers, particularly if situated in a seismically active region or if the underlying soils have geotechnical risk factors such as high compressibility or bay mud. They also pose serious challenges to firefighters during emergencies in high-rise structures. New and old building design, building systems like the building standpipe system, HVAC systems (heating, ventilation and air conditioning), fire sprinkler system and other things like stairwell and elevator evacuations pose significant problems. Studies are often required to ensure that pedestrian wind comfort and wind danger concerns are addressed. In order to allow less wind exposure, to transmit more daylight to the ground and to appear more slender, many high-rises have a design with setbacks.

Apartment buildings have technical and economic advantages in areas of high population density, and have become a distinctive feature of housing accommodation in virtually all densely populated urban areas around the world. In contrast with low-rise and single-family houses, apartment blocks accommodate more inhabitants per unit of area of land and decrease the cost of municipal infrastructure.

16-floor Rusland Court tower block in Charlestown, Greater Manchester, United Kingdom.

Definition

Various bodies have defined "high-rise":

- Emporis Standards defines a high-rise as "A multi-story structure between 35–100 meters tall, or a building of unknown height from 12–39 floors."
- According to the building code of Hyderabad, India, a high-rise building is one with four floors or more, or 15 to 18 meters or more in height.
- The *New Shorter Oxford English Dictionary* defines a high-rise as "a building having many storeys".
- The *International Conference on Fire Safety in High-Rise Buildings* defined a high-rise as "any structure where the height can have a serious impact on evacuation"
- In the U.S., the National Fire Protection Association defines a high-rise as being higher than 75 feet (23 meters), or about 7 stories.
- Most building engineers, inspectors, architects and similar professionals define a high-rise as a building that is at least 75 feet (23 m) tall.

History

High-rise apartment buildings had already appeared in antiquity: the *insulae* in ancient Rome and several other cities in the Roman Empire, some of which might have reached up to ten or more stories, one reportedly having 200 stairs. Because of the destruction caused by poorly built high-rise *insulae* collapsing, several Roman emperors, beginning with Augustus (r. 30 BC – 14 AD), set limits of 20–25 meters for multi-story buildings, but met with limited success, as these limits were often ignored despite the likelihood of taller *insulae* collapsing. The lower floors were typically occupied by either shops or wealthy families, while the upper stories were rented out to the lower classes. Surviving Oxyrhynchus Papyri indicate that seven-story buildings even existed in provincial towns, such as in third century AD Hermopolis in Roman Egypt.

These tower blocks were built in Shibam, Yemen in the 16th century, and are the tallest mudbrick buildings in the world.

In Arab Egypt, the initial capital city of Fustat housed many high-rise residential buildings, some seven stories tall that could reportedly accommodate hundreds of people. Al-Muqaddasi, in the 10th century, described them as resembling minarets, while Nasir Khusraw, in the early 11th century, described some of them rising up to 14 stories, with roof gardens on the top story complete with ox-drawn water wheels for irrigating them. By the 16th century, Cairo also had high-rise apartment buildings where the two lower floors were for commercial and storage purposes and the multiple stories above them were rented out to tenants.

The skyline of many important medieval cities was dominated by large numbers of high-rising urban towers, which fulfilled defensive but also representative purposes. The residential Towers of Bologna numbered between 80 and 100 at a time, the largest of which still rise to 97.2 m. In Florence, a law of 1251 decreed that all urban buildings should be reduced to a height of less than 26 m, the regulation immediately put into effect. Even medium-sized towns such as San Gimignano are known to have featured 72 towers up to 51 m in height.

The Hakka people in southern China have adopted communal living structures designed to be easily defensible in the forms of Weilongwu and Tulou, the latter are large, enclosed and fortified earth building, between three and five stories high and housing up to 80 families. The oldest still standing tulou dates back from the 14th century.

Tower blocks were built in the Yemeni city of Shibam in the 16th century. The houses of Shibam are all made out of mud bricks, but about five hundred of them are tower houses, which rise five to sixteen stories high, with each floor having one or two apartments. This technique of building was implemented to protect residents from Bedouin attacks. While Shibam has existed for around two thousand years, most of the city's houses date from the 16th century. The city has the tallest mud buildings in the world, some more than 30 meters (100 feet) high. Shibam has been called "one of the oldest and best examples of urban planning based on the principle of vertical construction" or "Manhattan of the desert".

Currently, the tallest high-rise apartment building in the world is Chicago's John Hancock Center, constructed under the supervision of Skidmore, Owings & Merrill and completed in 1969. The building has 100 stories and stands 344 meters tall.

Modern Development

A house in Steinfurt, Westphalia, Germany, forming a "Y".

The residential tower block with its typical concrete construction are a familiar feature of Modernist architecture. Influential examples include Le Corbusier's "housing unit" his *Unité d'Habitation*, repeated in various European cities starting with his *Cité radieuse* in Marseille (1947–52), constructed of *béton brut*, rough-cast concrete, as steel for framework was unavailable in post-war France. Residential tower blocks became standard in housing urban populations displaced by slum clearances and "urban renewal".

Highrise projects after World War II typically rejected the classical designs of the early skyscrapers, instead embracing the uniform international style; many older skyscrapers were redesigned to suit contemporary tastes or even got demolished - such as New York's Singer Building, once the world's tallest skyscraper. However, with the movements of Postmodernism, New Urbanism and New Classical Architecture, that established since the 1980s, a more classical approach came back to global skyscraper design, that is popular today.

Other contemporary styles and movements in highrise design include organic, sustainable, neo-futurist, structuralist, high-tech, deconstructivist, blob, digital, streamline, novelty, critical regionalist, vernacular, Neo Art Deco and neo-historist, also known as revivalist.

Europe

Eastern Europe

Winter in Nowa Huta (Kraków), Poland.

Typical 11-storey large-panel system building in Budapest-Kispest (built by the BHK III.)

Although some Eastern European countries during the interwar period, such as the Second Polish Republic, already started building housing estates that were considered to be of a very high standard for their time, many of these structures perished during the Second World War. In the Eastern Bloc, tower blocks were constructed in great numbers to produce plenty of cheap accommodation for the growing postwar populations of the USSR and its satellite states. This took place mostly in the 50s, 60s and 70s of the 20th century, though in the People's Republic of Poland this process started even earlier due to the severe damages that Polish cities sustained during World War II. Throughout the former Eastern Bloc countries, tower blocks built during the Soviet years make up much of the current housing estates and most of them were built in the specific socialist realist style of architecture that was dominant in the territories east of the Iron Curtain. In Romania, mass construction of standardized apartment blocks occurred under the systematization programme of Nicolae Ceauşescu, who, after a 1971 visit in North Korea, was impressed by the Juche ideology of that country. Systematization consisted largely of the demolition and reconstruction of existing villages, towns, and cities, in whole or in part, in order to build blocks of flats (*blocuri*). In the Czech Republic and Slovakia (then Czechoslovakia), panelák building under communism resulted from two main factors: the postwar housing shortage and the ideology of communist Czechoslovak leaders.

Painted paneláks in Prague, Czech Republic. In recent years, several Eastern European countries have started projects of colouring and beautifying existing communist blocks, so that they are no longer gray.

Nowadays, in Eastern European countries, opinions about these buildings vary greatly, with some deeming them as eyesores on their city's landscape while others glorify them as relics of a bygone age. Since the dissolution of the Soviet Union, and especially in the late 1990s and early 2000s, many of the former Eastern Bloc countries have begun construction of new, more expensive and modern housing.

The Śródmieście borough of Warsaw, the capital of Poland, has seen the development of an array of skyscrapers. Russia is also currently undergoing a dramatic buildout, growing a commercially shaped skyline. Moreover, the ongoing changes made to postwar housing estates in Eastern Europe since the 2000s in countries like Russia, Poland, parts of Germany, Czech Republic, and Estonia vary - ranging from simply applying a new coat of paint to the previously grey exterior to thorough modernization of entire buildings. In the European Union, among former communist countries, a majority of the population lives in flats in Latvia (65.1%), Estonia (63.8%), Lithuania (58.4%), Czech Republic (52.8%) and Slovakia (50.3%) (as of 2014, data from Eurostat). However, not all flat dwellers in Eastern Europe live in communist era blocks of flats; many live in buildings constructed after the fall of communism, and some in buildings surviving from the era before communism.

Great Britain

The three tower blocks of the Crossways Estate in Bow, London, United Kingdom, before their refurbishment.

Tower blocks were built in the UK after the Second World War. The first residential tower block, "The Lawn" was constructed in Harlow, Essex in 1951; it is now a Grade II listed building. In many cases tower blocks were seen as a "quick-fix" to cure problems caused by crumbling and unsanitary 19th-century dwellings or to replace buildings destroyed by German aerial bombing. It was argued that towers surrounded by public open space could provide the same population density as the terraced housing and small private gardens they replaced, offering larger rooms and improved views whilst being cheaper to build. Initially, they were welcomed, and their excellent views made them popular living places. Later, as the buildings themselves deteriorated, they grew a reputation for being undesirable low cost housing, and many tower blocks saw rising crime levels, increasing their unpopularity. One response to this was the great increase in the number of housing estates built, which in turn brings its own problems. In the UK, tower blocks particularly lost popularity after the partial collapse of Ronan Point in east London in 1968.

Glasgow, the largest city in Scotland, is believed to contain the highest concentration of tower blocks in the UK: examples include the Hutchesontown C blocks in the Gorbals and the 20-storey blocks in Sighthill. However, on the whole, London has the largest number of high-rise residential buildings in the UK.

The Post-war British Tower Block Vision

Post-war Britain was the stage for a tower block "building boom"; from the 1950s to the late 1970s there was a dramatic increase in tower block construction. During this time, local authorities desired to impress their voters by building futuristic and imposing tower blocks, which would signify post-war progress. Both Patrick Dunleavy and Lynsey Hanley agree that architects and planners were influenced by Le Corbusier's promotion of high-rise architecture. The modern tower blocks were to include features that would foster desired forms of resident interaction, an example being the inclusion of Le Corbusier's streets in the sky in some estates.

As well as inspiring residents, local authority planners believed that the way tower blocks were constructed would save money. Generally, the tower blocks were built on cheap greenfield land skirting established cities. Although the property prices for these periphery sites were markedly cheaper than their inner city counterparts, they often had little access to public amenities, such as public transport. It was thought that the implementation of industrialised building techniques would lower costs too, as similar tower blocks would be replicated over many sites. Uniform and standardised parts, such as toilet fittings and door handles, would be fitted throughout many tower blocks; planners deemed that buying in bulk would reduce overall costs.

Another key aspect of the tower block vision was the Brutalist architectural method, popular with architects and planners at the time. The Brutalist emphasis led to the construction of stark and striking tower blocks with large sections of exposed concrete. Concrete was to be an integral part of the tower block designs; it could be poured on site, offering boundless flexibility to the building designers. To the planners, concrete was a silver bullet for the construction process – it was economical, and "was vaunted as being long-lasting, if not indestructible".

The Post-war British Tower Block Reality

Elmet Towers in Swarcliffe, Leeds, showing the dereliction that led to its demolition.

Coleman's 1985 work argues that in trying to emulate Le Corbusier's ideas, the tower block planners only succeeded in encouraging social problems. Although architects and local authorities intended the opposite, tower blocks quickly became, as Hanley sharply stated, 'slums in the sky'. Due to demanding deadlines, complicated construction practices were rushed and many tower blocks experienced structural decay as a result – roofs leaked, concrete suffered spalling, steel corroded, and damp penetrated the buildings. Unfortunately, by replicating tower blocks across the nation, planners 'disastrously' replicated design faults. In many tower blocks, concrete quickly exhibited signs of decay; cracks soon formed and destabilised the buildings. The partial collapse of the Ronan Point tower block is an infamous example of the hasty and substandard construction that occurred in a number of the towers. The tower blocks quickly lost their "futuristic" look; concrete turned from the crisp white the designers had imagined to a dull grey, stained by pollution.

Poor design decisions ruined the anticipated benefits of the buildings. Open spaces, which were supposed to benefit the residents, were instead unattractive, unused and inadequately supervised. Residents felt it was difficult to maintain the large open spaces around the blocks because they realistically belonged to no one. Social problems increased as the tower blocks quickly degraded because of poor maintenance and an insecure communal environment. Apart from frequent break-downs, communal lifts were a source of fear for people travelling alone. It was a rarity to "enter a clean-smelling, undefaced lift". The tower blocks, many of which were on the periphery of the city, made residents feel isolated and cut off from society. Outsiders and newcomers were also affected; they felt the overbearing design of the tower blocks made them fearsome and unsociable.

Power argues that as a direct consequence of their design and construction, security problems were prevalent in many of the tower blocks. Break-ins, vandalism and muggings were common, which were aided by the buildings' concealed areas, the mazes of internal corridors, and dark corners. Police were often required in the tower blocks, but their infrequent presence did little to pacify towers rife with delinquency. To contain disruptive behaviour, local authorities began to place "problem families" in the same blocks; Hanley argues that this policy only led to "further alienation ... nihilism and a creeping sense of lawlessness". Dunleavy seconds this, suggesting that the mental health of long-term tower block residents may have been detrimentally affected.

While local authorities and their architects intended to create tower blocks that encouraged harmonious and vibrant communities, often the results were far from ideal. Post-war tower blocks were compromised from the outset by a combination of faults: local authorities advocated impractical architectural methods; design and construction faults were frequently reproduced; and there appeared to be a lack of understanding about the social consequences of certain design features. Collectively, these oversights transformed many tower blocks into undesirable places to live.

High-rise apartment buildings of the "Komplex Leipziger Straße" in Berlin-Mitte.

Fire Safety

Compliant protections for residents in tower blocks can be very expensive to retrofit, and lessors (such as local authorities) have been sued for lack of a current fire safety inspection, or implementation of its recommendations. Ed Davey MP commented regarding a life-claiming 2009 fire that many blocks in the UK remained inadequate. North Ayrshire Council has decided to remove plastic cladding from its buildings following a fatal fire in a tower block. Poor maintenance of electrical equipment, fire doors and other features has caused widespread fire hazards and residents are asked to pay part of the cost of putting this right even though the problem is due to poor maintenance and errors in design.

Tower blocks may be inherently more prone to casualties from a fire because people living on higher floors cannot escape fires easily and the fire brigade cannot reach the higher floors quickly. In buildings with more than a hundred residents, ensuring that every single resident acts responsibly to minimize fire risk is difficult; poorer residents in tower blocks may be tempted to use cheaper flammable fuels rather than electricity, they are also more likely to be smokers (carelessness with cigarettes is a major cause of home fires), and they are more likely to have old furniture, not made to modern fire safety standards. Fire safety legislation introduced in 2006 requires new high rise buildings to be built to higher safety standards with sprinkler systems; the same standards do not apply to pre-2006 tower blocks, which contain a greater proportion of poor people. Recent studies have investigated the combined use of egress components (e.g., stairs and elevators) to enhance the effectiveness of evacuation strategies in case of fire.

Towards the Present Day

The 31-story Red Road estate in Glasgow, the tallest public housing towers in Europe at the time of their construction. Like many 1960s British high-rises, they are now condemned.

In recent years, some council or ex-council high-rises in the United Kingdom, including Trellick Tower, Keeling House, Sivill House and The Barbican Estate, have become popular with young professionals due to their excellent views, desirable locations and architectural pedigrees, and now command high prices. There are plans to redevelop the Little London and Lovell Park areas on the fringes of Leeds city centre into luxury flats for 'Young Urban Professionals'. The plans entail demolishing all of the council housing and refurbishing the highrise flats. This demand has led to many councils rethinking plans regarding their demolition.

In Glasgow, similar initiatives have taken place – in 2011, the Glasgow Housing Association saved one of a cluster of three condemned tower blocks in the Ibrox area with the help of government grants to transform it into desirable rented housing for young professionals. Glasgow itself has taken a more measured approach to its high-rise housing stock, eschewing the mass demolition programmes being practiced in other British cities – favouring selective refurbishment of estates with secure futures and on buildings in good structural condition, only razing towers where absolutely necessary.

After a gap of around 30 years, new high-rise flats are once again being built in Belfast, Birmingham (some are for wealthy people, e.g. The Mailbox, others are student accommodation), Cardiff, Aberdeen, Glasgow, Leeds, Liverpool, London, Manchester, and Newcastle upon Tyne; but this time sometimes for wealthy professionals, rather than the "lower classes." Their developers market these properties by using the American term 'apartment buildings', perhaps in an effort to distance these newer buildings from the older tower blocks from the 1950s and 1960s. These are usually taller than their older counterparts and generally built in and around these provincial city centres. They are often glass and aluminium clad. Tonight with Trevor McDonald highlighted that in Leeds and Manchester (perhaps the cities that had seen most development) only approximately half were occupied and with owner occupation often being as low as 10%. In Southend there are controversial plans to build new tower blocks for social housing, a supporter claims this will be cheaper than paying to house poor people with private landlords while opponents fear the creation of Ghettos.

Ireland

Flats in Ballymun, Dublin, Ireland.

Republic of Ireland

In Dublin, the Ballymun Flats were built between 1966 and 1969: seven 15-story towers, nineteen 8-story blocks and ten 4-story blocks. These were the "seven towers" referred to in the U2 song

"Running to Stand Still". They have since been demolished. Inner Dublin flat complexes include Sheriff Street (demolished), Fatima Mansions (demolished and redeveloped), St. Joseph's Gardens (demolished; replaced by Killarney Court flat complex), St. Teresa's Gardens, Dolphin House, Liberty House, St. Michael's Estate and O'Devaney Gardens and a lot more mainly throughout the North and South Inner City of Dublin. Suburban flat complexes were built exclusively on the northside of the city in Ballymun, Coolock and Kilbarrack. These flats were badly affected by a heroin epidemic that hit working-class areas of Dublin in the 1980s and early 90s.

Over the last five years the largest cities such as Dublin, Cork, Limerick and Galway have witnessed new large apartment building. Some large towns such as Navan, Drogheda, Dundalk and Mullingar have also witnessed lots of modern apartments being built.

Northern Ireland

Tower blocks in Northern Ireland were never built to the frequency as in cities on the island of Great Britain. Most tower blocks and flat complexes are found in Belfast although many of these have been demolished since the 1990s and replaced with traditional public housing units. The mid-rise Divis flats complex in west Belfast was built in between 1968 and 1972. It was demolished in the early 1990s as the residents demanded new houses due to mounting problems with the flats. Divis Tower, built separately in 1966, still stands, however; and in 2007 work began to convert the former British Army base at the top two floors into new dwellings. Divis Tower was for several decades Ireland's tallest residential building, having since being surpassed by the privately owned Obel Tower in the city centre. In the north of the City, the iconic 7 towers complex in the New Lodge remains, although so too the problems that residents face, such as poor piping and limited sanitation. Farther north, the four tower blocks in Rathcoole dominate the local skyline, while in south Belfast, the tower blocks in Seymour Hill, Belvoir & Finaghy remain standing.

Most of the aforementioned high rise flats in the city were built by the Northern Ireland Housing Trust (NIHT) as part of overspill housing schemes, the first such development being the pair of point blocks in East Belfast's Cregagh estate. These eleven-story towers were completed in 1961 and were the first tall council housing blocks on the island of Ireland. The NIHT also designed the inner city Divis Flats complex. The six to eight story deck access flats that comprised most of the Divis estate were of poor build quality and were all demolished by the early 1990s. Similar slab blocks were built by the NIHT in East Belfast (Tullycarnet) and Derry's Bogside area, all four of which have been demolished.

Belfast Corporation constructed seven tower blocks on the former Victoria Barracks site in the New Lodge district. While the Corporation built some mid-rise flats as part of slum clearance schemes (most notably the now demolished Unity Flats and the 'Weetabix Flats in the Shankill area), New Lodge would be their only high rise project in the inner city with three more in outlying areas of the city during the 1960s, two being in Mount Vernon in North Belfast and one being in the Clarawood estate, East Belfast. The Royal Hospital built three thirteen-story towers for use as staff accommodation, prominently located adjacent to the M2 Motorway at Broadway. Belfast City Hospital also constructed a high rise slab block which since privatisation has been named Bradbury Court, formerly known as Erskine House. Queens University Belfast built several eleven storey towers at their Queens Elms student accommodation. Of the three sixteen-story point blocks of Larne Borough Council in the late 1960s, only one remains.

Asia

Residential tower complexes are common in Asian countries such as China, India, Indonesia, Taiwan, Singapore, Japan, and South Korea as urban densities are very high. In Singapore and urban Hong Kong, land prices are so high that a large portion of the population lives in high-rise apartments. In fact, over 60% of Hong Kong residents live in apartments, many of them condominiums.

Sarah Williams Goldhagen (2012) celebrated the work of innovative architecture firms such as WOHA (based in Singapore), Mass Studies (based in Seoul), Amateur Architecture Studio (based in Hangzhou), and the New York City-based Steven Holl in the transformation of residential towers into "vertical communities" or "vertical cities in the sky" providing aesthetic, unusually designed silhouettes on the skyline, comfortable private spaces and attractive public spaces. None of these "functional, handsome, and humane high-rise residential buildings" are affordable housing.

South Korea

Typical Tower block apartment in South Korea.

In South Korea the tower blocks are called Apartment Complex. The first residential towers began to be built after the Korean War. The South Korean government needed to build many apartment complexes in the cities to be able to accommodate the citizens. In the 60 years since, as the population increased considerably, tower blocks have become more common. This time however the new tower blocks integrated shopping malls, parking systems and other convenient facilities.

Samsung Tower Palace in Seoul, South Korea, is the tallest apartment complex in Asia.

In Seoul, approximately 80% of its residents live in apartment complexes which comprise 98% of recent residential construction. Seoul proper is noted for its population density, almost twice that of New York and eight times greater than Rome, though slightly less than Paris. Its metropolitan area is the densest in the OECD.

China

The 2012 Pritzker Prize was awarded to Chinese architect Wang Shu. Among his winning designs is the Vertical Courtyard Apartments, six 26-story towers built in Hangzhou, China, by his architectural firm, Amateur Architecture Studio. "These towers were designed to house two-storey apartments, in which every inhabitant would enjoy "the illusion of living on the second floor", accomplished by folding concrete floor planes (like "bamboo mats," claims the firm), so that every third story opens into a private courtyard. In the larger towers, the two-story units are stacked slightly askew, adding to the visual interest of the variegated façades (Goldhagen 2012)."

Americas

Residential buildings in Santa Fe, Mexico City, Mexico.

Canada

In Canada large multi-family buildings are usually known as *apartment buildings* or *apartment blocks* if they are rented from one common landowner, or *condominiums* or *condo towers* if each dwelling unit is individually owned; they may be called *low-rise* (or *walk-up*), *mid-rise*, *high-rise*, or *skyscraper* depending on their height. Tall residential towers are a staple building type in all large cities. Their relative prominence in Canadian cities varies substantially, however. In general, more populated cities have more high-rises than smaller cities, due to a relative the scarcity of land and a greater demand for housing.

However, some cities such as Quebec City, and Halifax have fewer high-rise buildings due to several factors: a focus on historic preservation, height restrictions, and lower growth rates. In middle-sized cities with a relatively low population density, such as Calgary, Edmonton, Winnipeg, or Hamilton, there are more apartment towers but they are greatly outnumbered by single-family houses. Most of the largest residential towers in Canada are found in Montreal, Toronto, and Vancouver—the country's most densely populated cities.

Toronto contains the second largest concentration of high-rise apartment buildings in North America (after New York). In Canada, like in other New World countries, but unlike Western Europe, most high-rise towers are located in the city centre (or "downtown"), where smaller, older buildings were demolished to make way in redevelopment schemes.

United States

432 Park Avenue, the tallest residential high-rise tower in the world, under construction in July 2014.

In the United States tower blocks are commonly referred to as "midrise" or "highrise apartment buildings", depending on their height, while buildings that house fewer flats (apartments), or are not as tall as the tower blocks, are called "lowrise apartment buildings". Specifically, "midrise" buildings are as tall as the streets are wide, allowing 5 hours of sunlight on the street.

Some of the first residential towers were the Castle Village towers in Manhattan, New York City, completed in 1939. Their cross-shaped design was copied in towers in Parkchester and Stuyvesant Town residential developments.

The government's experiments in the 1960s and 70s to use high-rise apartments as a means of providing the housing solution for the poor broadly resulted in failure. Made in the tower in the park style, all but a few high-rise housing projects in the nation's largest cities, such as Cabrini–Green and Robert Taylor Homes in Chicago, Penn South in New York City and the Desire projects in New Orleans, fell victim to the "ghettofication" and are now being torn down, renovated, or replaced. Another example is the former Pruitt–Igoe complex in St. Louis, torn down in the 1970s.

In contrast to their public housing cousins, commercially developed high-rise apartment buildings continue to flourish in cities around the country largely due to high land prices and the housing boom of the 2000s. The Upper East Side in New York City and Chicago's Gold Coast, both featuring high-rise apartments, are the wealthiest urban neighborhoods in the United States.

Currently, the tallest residential building in the world is 432 Park Avenue located in New York City, United States of America, having a height of 1396 feet with the highest occupied floor at 1287 feet.

Australia

High-rise living in Australia was limited to small pockets of bohemian inner Sydney until the 1960s, where a short-lived fashion saw public housing tenants located in new high-rise developments, especially in Sydney and Melbourne. The buildings pictured along with four other 16-story blocks were constructed on behalf of the Royal Australian Navy and were available to sailors and their families for accommodation. Due to social problems within these blocks the Navy left and the Department of Housing took charge and flats were let to low income and immigrant families. During the 1980s many people escaping communism in Eastern block countries were housed in these buildings. Developers have enthusiastically adopted the term "apartment" for these new high-rise blocks, perhaps to avoid the stigma still attached to housing commission flats.

Housing commission towers in Waterloo, Sydney, Australia.

Streets in the Sky

"Street in the sky" at Park Hill.

Streets in the sky is a style of architecture that emerged in Britain in the 1960s and 1970s. Generally built to replace run-down terraced housing, the new designs included not only modern improvements such as inside toilets, but also shops and other community facilities within high-rise blocks. Examples of the buildings and developments are Trellick Tower, Balfron Tower, Robin Hood Gardens and Keeling House in London, Hunslet Grange in Leeds and Park Hill, Sheffield. These were an attempt to develop a new architecture, differentiated from earlier large housing estates, such as Quarry Hill flats in Leeds. Alison and Peter Smithson were the architects of Robin Hood Gardens. Another large example, the Aylesbury Estate in South London, built in 1970, is about to be demolished. The Hulme Crescents in Manchester were the largest social housing scheme in Europe when built in 1972 but lasted just 22 years. The Crescents are recognised as one of the worst social housing schemes in British history, marred by design and practical problems; they severely challenged the viability of the Streets in the Sky ideal.

Deck Access

Deck access is a type of flat that is accessed from a walkway that is open to the elements, as opposed to flats that are accessed from fully enclosed internal corridors. Deck access blocks of flats are usually fairly low-rise structures. The decks can vary from simple walkways, which may be covered or uncovered, to decks wide enough for small vehicles. The best known example of deck-access flats in the UK is Park Hill, Sheffield, where the decks are wide enough to allow electric vehicles; however, the design is inspired by French Modernist architect Le Corbusier, particularly his Unite D'Habitation in Marseilles.

Green Tower Blocks

Salford tower blocks in 2001. In the UK, tower blocks were mostly built between the 1950s and 70s.

Green tower blocks have some scheme of living plants or green roofs or solar panels on their roofs and/or incorporate other environmentally friendly design features.

Architectural and Structural Engineering

César Pelli's Ratner Athletic Center uses cables and masts as load-bearing devices.

Architectural engineering, also known as building engineering, is the application of engineering principles and technology to building design and construction. Definitions of an architectural engineer may refer to:

- An engineer in the structural, mechanical, electrical, construction or other engineering fields of building design and construction.
- A licensed engineering professional in parts of the United States.
- Architectural engineers are those who work with other engineers and architects for the designing and construction of buildings.

Engineering for Designer

Structural Engineering

Structural engineering involves the analysis and design of physical objects (buildings, bridges, equipment supports, towers and walls). Those concentrating on buildings are responsible for the structural performance of a large part of the built environment and are, sometimes, informally referred to as "building engineers". Structural engineers require expertise in strength of materials and in the seismic design of structures covered by earthquake engineering. Architectural Engineers sometimes structural as one aspect of their designs; the structural discipline when practiced as a specialty works closely with architects and other engineering specialists.

Mechanical, Electrical, and Plumbing (MEP)

MEP room in a building.

Mechanical engineering and electrical engineering engineers are specialists, commonly referred to as (mechanical, electrical, and plumbing) when engaged in the building design fields. Also known as "building services engineering" in the United Kingdom, Canada, and Australia. Mechanical engineers often design and oversee the heating, ventilation and air conditioning (HVAC), plumbing, and rain gutter systems. Plumbing designers often include design specifications for simple active fire protection systems, but for more complicated projects, fire protection engineers are often separately retained. Electrical engineers are responsible for the building's power distribution, telecommunication, fire alarm, signalization, lightning protection and control systems, as well as lighting systems.

The Architectural Engineer (PE) in the United States

In many jurisdictions of the United States, the architectural engineer is a licensed engineering professional. Usually a graduate of an architectural engineering university program preparing students to perform whole-building design in competition with architect-engineer teams; or for practice in one of structural, mechanical or electrical fields of building design, but with an appreciation of integrated architectural requirements.

Formal architectural engineering education, following the engineering model of earlier disciplines, developed in the late 19th century, and became widespread in the United States by the mid-20th century. With the establishment of a specific "architectural engineering" NCEES Professional Engineering registration examination in the 1990s, and first offering in April 2003, architectural engineering became recognized as a distinct engineering discipline in the United States. Architectural engineers are not entitled to practice architecture unless they are also licensed as architects.

The Architect as Architectural Engineer

In some countries, the practice of architecture includes planning, designing and overseeing the building's construction, and architecture, as a profession providing architectural services, is referred to as "architectural engineering". In Japan, a "first-class architect" plays the dual role of architect and building engineer, although the services of a licensed "structural design first-class architect" are required for buildings over a certain scale.

In some languages, such as Korean and Arabic, "architect" is literally translated as "architectural engineer". In some countries, an "architectural engineer" (such as the *ingegnere edile* in Italy) is entitled to practice architecture and is often referred to as an architect. These individuals are often also structural engineers. In other countries, such as Germany, Austria, Iran, and most of the Arab countries, architecture graduates receive an engineering degree (*Dipl.-Ing. – Diplom-Ingenieur*).

In Spain, an "architect" has a technical university education and legal powers to carry out building structure and facility projects.

In Brazil, architects and engineers used to share the same accreditation process (CONFEA – Federal Council of Engineering, Architecture and Agronomy). Now the Brazilian architects and urbanists have their own accreditation process (CAU – Architecture and Urbanism Council). Besides traditional architecture design training, Brazilian architecture courses also offer complementary training in engineering disciplines such as structural, electrical, hydraulic and mechanical engineering. After graduation, architects focus in architectural planning, yet they can be responsible to the whole building, when it concerns to small buildings (except in electric wiring, where the architect autonomy is limited to systems up to 30kVA, and it has to be done by an Electrical Engineer), applied to buildings, urban environment, built cultural heritage, landscape planning, interiorscape planning and regional planning.

In Greece licensed architectural engineers are graduates from architecture faculties that belong to the Polytechnic University, obtaining an "Engineering Diploma". They graduate after 5 years of studies and are fully entitled architects once they become members of the Technical Chamber of Greece (TEE – Τεχνικό Επιμελητήριο Ελλάδος). The Engineering Diploma equals a master's degree in ECTS units (300) according to the Bologna Accords.

Education

The architectural, structural, mechanical and electrical engineering branches each have well established educational requirements that are usually fulfilled by completion of a university program.

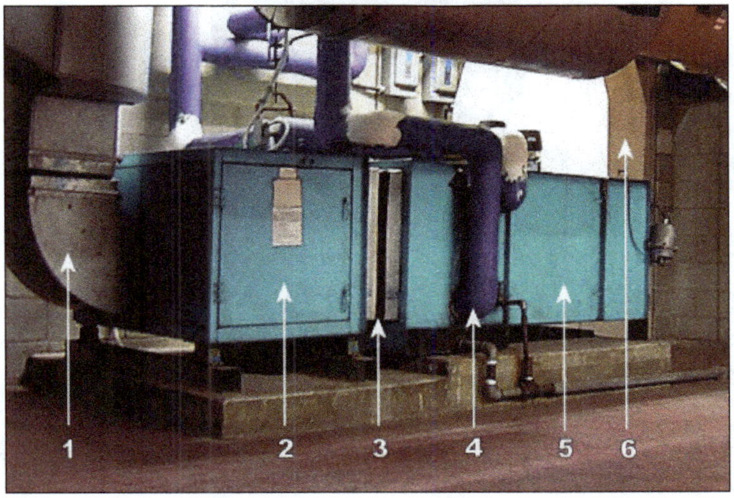

An air handling unit is used for the heating and cooling of air in a central location (click on image for legend). Bringing together knowledge of acoustic engineering and HVAC is one example of the multi-disciplined nature of architectural engineering.

Architectural Engineering as a Single Integrated Field of Study

What differentiates architectural engineering from architecture (architect) as a separate and single, integrated field of study, compared to other engineering disciplines, is its multi-disciplinary engineering approach. Through training in and appreciation of architecture, the field seeks integration of building systems within its overall building design. Architectural engineering includes the design of building systems including heating, ventilation and air conditioning (HVAC), plumbing, fire protection, electrical, lighting, architectural acoustics, and structural systems. In some university programs, students are required to concentrate on one of the systems; in others, they can receive a generalist architectural or building engineering degree.

Structural Engineering

Structural engineering is mainly a sub-division of civil engineering in which structural engineers are trained to understand, predict, and calculate the stability, strength and rigidity of built structures for buildings and nonbuilding structures, to develop designs and integrate their design with that of other designers, and to supervise construction of projects on site. They can also be involved in the design of machinery, medical equipment, and vehicles where structural integrity affects functioning and safety.

Structural engineering theory is based upon applied physical laws and empirical knowledge of the structural performance of different materials and geometries. Structural engineering design utilizes a number of relatively simple structural elements to build complex structural systems. Structural engineers are responsible for making creative and efficient use of funds, structural elements and materials to achieve these goals.

The Eiffel Tower in Paris is a historical achievement of structural engineering.

Structural Engineer (Professional)

Structural engineers are responsible for engineering design and analysis. Entry-level structural engineers may design the individual structural elements of a structure, for example the beams, columns, and floors of a building. More experienced engineers may be responsible for the structural design and integrity of an entire system, such as a building.

Structural engineers often specialize in particular fields, such as bridge engineering, building engineering, pipeline engineering, industrial structures, or special mechanical structures such as vehicles, ships or aircraft.

Structural engineering has existed since humans first started to construct their own structures. It became a more defined and formalized profession with the emergence of the architecture as distinct profession from the engineering during the industrial revolution in the late 19th century. Until then, the architect and the structural engineer were usually one and the same thing - the master builder. Only with the development of specialized knowledge of structural theories that emerged during the 19th and early 20th centuries, did the professional structural engineers come into existence.

The role of a structural engineer today involves a significant understanding of both static and dynamic loading, and the structures that are available to resist them. The complexity of modern structures often requires a great deal of creativity from the engineer in order to ensure the structures support and resist the loads they are subjected to. A structural engineer will typically have a four or five year undergraduate degree, followed by a minimum of three years of professional practice before being considered fully qualified. Structural engineers are licensed or accredited by different learned societies and regulatory bodies around the world (for example, the Institution of Structural Engineers in the UK). Depending on the degree course they have studied and/or the jurisdiction they are seeking licensure in, they may be accredited (or licensed) as just structural engineers, or as civil engineers, or as both civil and structural engineers. Another international organisation is IABSE (International Association for Bridge and Structural Engineering). The aim of that association is to exchange knowledge and to advance the practice of structural engineering worldwide in the service of the profession and society.

History of Structural Engineering

Pont du Gard, France, a Roman era aqueduct circa 19 BC.

Structural engineering dates back to 2700 B.C.E. when the step pyramid for Pharaoh Djoser was built by Imhotep, the first engineer in history known by name. Pyramids were the most common major structures built by ancient civilizations because the structural form of a pyramid is inherently stable and can be almost infinitely scaled (as opposed to most other structural forms, which cannot be linearly increased in size in proportion to increased loads).

The structural stability of the pyramid is not primarily a result of its shape, as the integrity of the pyramid is intact as long as each of the stones is able to support the weight of the stone above it. The limestone blocks were taken from a quarry near the build site. Since the compressive strength of limestone is anywhere from 30 to 250 MPa (MPa = Pa * 10^6), the blocks will not fail under compression. Therefore, the structural strength of the pyramid stems from the material properties of the stones from which it was built rather than the pyramid's geometry.

Throughout ancient and medieval history most architectural design and construction was carried out by artisans, such as stone masons and carpenters, rising to the role of master builder. No theory of structures existed, and understanding of how structures stood up was extremely limited, and based almost entirely on empirical evidence of 'what had worked before'. Knowledge was retained by guilds and seldom supplanted by advances. Structures were repetitive, and increases in scale were incremental.

No record exists of the first calculations of the strength of structural members or the behavior of structural material, but the profession of structural engineer only really took shape with the Industrial Revolution and the re-invention of concrete. The physical sciences underlying structural engineering began to be understood in the Renaissance and have since developed into computer-based applications pioneered in the 1970s.

Timeline

- 1452–1519 Leonardo da Vinci made many contributions.
- 1638: Galileo Galilei published the book *Two New Sciences* in which he examined the failure of simple structures.

Galileo Galilei published the book *Two New Sciences* in which he examined the failure of simple structures.

- 1660: Hooke's law by Robert Hooke.
- 1687: Isaac Newton published *Philosophiae Naturalis Principia Mathematica* which contains the Newton's laws of motion.

Isaac Newton published *Philosophiae Naturalis Principia Mathematica* which contains the Newton's laws of motion.

- 1750: Euler–Bernoulli beam equation.
- 1700–1782: Daniel Bernoulli introduced the principle of virtual work.
- 1707–1783: Leonhard Euler developed the theory of buckling of columns.

Leonhard Euler developed the theory of buckling of columns.

- 1826: Claude-Louis Navier published a treatise on the elastic behaviors of structures.
- 1873: Carlo Alberto Castigliano presented his dissertation "Intorno ai sistemi elastici", which contains his theorem for computing displacement as partial derivative of the strain energy. This theorem includes the method of "least work" as a special case.
- 1874: Otto Mohr formalized the idea of a statically indeterminate structure.
- 1922: Timoshenko corrects the Euler-Bernoulli beam equation.
- 1936: Hardy Cross' publication of the moment distribution method, an important innovation in the design of continuous frames.
- 1941: Alexander Hrennikoff solved the discretization of plane elasticity problems using a lattice framework.
- 1942: R. Courant divided a domain into finite subregions.
- 1956: J. Turner, R. W. Clough, H. C. Martin, and L. J. Topp's paper on the "Stiffness and Deflection of Complex Structures" introduces the name "finite-element method" and is widely recognized as the first comprehensive treatment of the method as it is known today.

Structural Failure

The history of structural engineering contains many collapses and failures. Sometimes this is due to obvious negligence, as in the case of the Pétion-Ville school collapse, in which Rev. Fortin Augustin *"constructed the building all by himself, saying he didn't need an engineer as he had good knowledge of construction"* following a partial collapse of the three-story schoolhouse that sent neighbors fleeing. The final collapse killed 94 people, mostly children.

In other cases structural failures require careful study, and the results of these inquiries have resulted in improved practices and greater understanding of the science of structural engineering. Some such studies are the result of forensic engineering investigations where the original engineer seems to have done everything in accordance with the state of the profession and acceptable practice yet a failure still eventuated. A famous case of structural knowledge and practice being advanced in this manner can be found in a series of failures involving box girders which collapsed in Australia during the 1970s.

Specializations

Building Structures

Sydney Opera House, designed by Architect Jørn Utzon and structural design by Ove Arup & Partners.

Structural building engineering includes all structural engineering related to the design of buildings. It is a branch of structural engineering closely affiliated with architecture.

Structural building engineering is primarily driven by the creative manipulation of materials and forms and the underlying mathematical and scientific ideas to achieve an end which fulfills its functional requirements and is structurally safe when subjected to all the loads it could reasonably be expected to experience. This is subtly different from architectural design, which is driven by the creative manipulation of materials and forms, mass, space, volume, texture and light to achieve an end which is aesthetic, functional and often artistic.

Millennium Dome in London, UK, by Richard Rogers and Buro Happold.

The architect is usually the lead designer on buildings, with a structural engineer employed as a sub-consultant. The degree to which each discipline actually leads the design depends heavily on the type of structure. Many structures are structurally simple and led by architecture, such as multi-storey office buildings and housing, while other structures, such as tensile structures, shells and gridshells are heavily dependent on their form for their strength, and the engineer may have a more significant influence on the form, and hence much of the aesthetic, than the architect.

Burj Khalifa, in Dubai, the world's tallest building, shown under construction in 2007 (since completed).

The structural design for a building must ensure that the building is able to stand up safely, able to function without excessive deflections or movements which may cause fatigue of structural elements, cracking or failure of fixtures, fittings or partitions, or discomfort for occupants. It must account for movements and forces due to temperature, creep, cracking and imposed loads. It must also ensure that the design is practically buildable within acceptable manufacturing tolerances of the materials. It must allow the architecture to work, and the building services to fit within the building and function (air conditioning, ventilation, smoke extract, electrics, lighting etc.). The structural design of a modern building can be extremely complex, and often requires a large team to complete.

Structural engineering specialties for buildings include:

- Earthquake engineering
- Façade engineering
- Fire engineering
- Roof engineering
- Tower engineering
- Wind engineering

Earthquake Engineering Structures

Earthquake engineering structures are those engineered to withstand earthquakes.

The main objectives of earthquake engineering are to understand the interaction of structures with the shaking ground, foresee the consequences of possible earthquakes, and design and construct the structures to perform during an earthquake.

Earthquake-proof pyramid El Castillo, Chichen Itza.

Earthquake-proof structures are not necessarily extremely strong like the El Castillo pyramid at Chichen Itza shown above. In fact, many structures considered strong may in fact be stiff, which can result in poor seismic performance.

One important tool of earthquake engineering is base isolation, which allows the base of a structure to move freely with the ground.

Civil Engineering Structures

Civil structural engineering includes all structural engineering related to the built environment. It includes:

• Bridges • Dams • Earthworks • Foundations • Offshore structures • Pipelines		• Power stations • Railways • Retaining structures and walls • Roads • Tunnels • Waterways • reservoir • Water and wastewater infrastructure

The structural engineer is the lead designer on these structures, and often the sole designer. In the design of structures such as these, structural safety is of paramount importance (in the UK, designs for dams, nuclear power stations and bridges must be signed off by a chartered engineer).

Civil engineering structures are often subjected to very extreme forces, such as large variations in temperature, dynamic loads such as waves or traffic, or high pressures from water or compressed gases. They are also often constructed in corrosive environments, such as at sea, in industrial facilities or below ground.

Mechanical Structures

Car

Principles of structural engineering are applied to variety of mechanical (moveable) structures. The design of static structures assumes they always have the same geometry (in fact, so-called static structures can move significantly, and structural engineering design must take this into account where necessary), but the design of moveable or moving structures must account for fatigue, variation in the method in which load is resisted and significant deflections of structures.

Motorbike

The forces which parts of a machine are subjected to can vary significantly, and can do so at a great rate. The forces which a boat or aircraft are subjected to vary enormously and will do so thousands of times over the structure's lifetime. The structural design must ensure that such structures are able to endure such loading for their entire design life without failing.

These works can require mechanical structural engineering:

- Boilers and pressure vessels
- Coachworks and carriages
- Cranes
- Elevators
- Escalators
- Marine vessels and hulls

Aerospace Structures

An Airbus A380, the world's largest passenger airliner.

Aerospace structure types include launch vehicles, (Atlas, Delta, Titan), missiles (ALCM, Harpoon), Hypersonic vehicles (Space Shuttle), military aircraft (F-16, F-18) and commercial aircraft (Boeing 777, MD-11). Aerospace structures typically consist of thin plates with stiffeners for the external surfaces, bulkheads and frames to support the shape and fasteners such as welds, rivets, screws and bolts to hold the components together.

Design of missile needs in depth understanding of Structural Analysis.

Nanoscale Structures

A nanostructure is an object of intermediate size between molecular and microscopic (micrometer-sized) structures. In describing nanostructures it is necessary to differentiate between the number of dimensions on the nanoscale. Nanotextured surfaces have one dimension on the nanoscale, i.e., only the thickness of the surface of an object is between 0.1 and 100 nm. Nanotubes have two dimensions on the nanoscale, i.e., the diameter of the tube is between 0.1 and 100 nm; its length could be much greater. Finally, spherical nanoparticles have three dimensions on the nanoscale, i.e., the particle is between 0.1 and 100 nm in each spatial dimension. The terms nanoparticles and ultrafine particles (UFP) often are used synonymously although UFP can reach into the micrometre range. The term 'nanostructure' is often used when referring to magnetic technology.

Structural Engineering for Medical Science

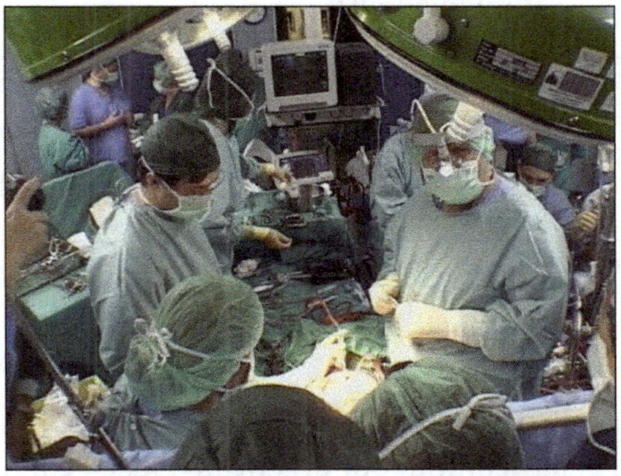

Designing Medical Equipment needs in-depth understanding of Structural Engineering.

Medical equipment (also known as armamentarium) is designed to aid in the diagnosis, monitoring or treatment of medical conditions. There are several basic types: Diagnostic equipment includes medical imaging machines, used to aid in diagnosis ; equipment includes infusion pumps, medical lasers and LASIK surgical machines ; Medical monitors allow medical staff to measure a patient's medical state. Monitors may measure patient vital signs and other parameters including ECG, EEG, blood pressure, and dissolved gases in the blood ; Diagnostic Medical Equipment may also be used in the home for certain purposes, e.g. for the control of diabetes mellitus. A biomedical equipment technician (BMET) is a vital component of the healthcare delivery system. Employed primarily by hospitals, BMETs are the people responsible for maintaining a facility's medical equipment.

Structural Elements

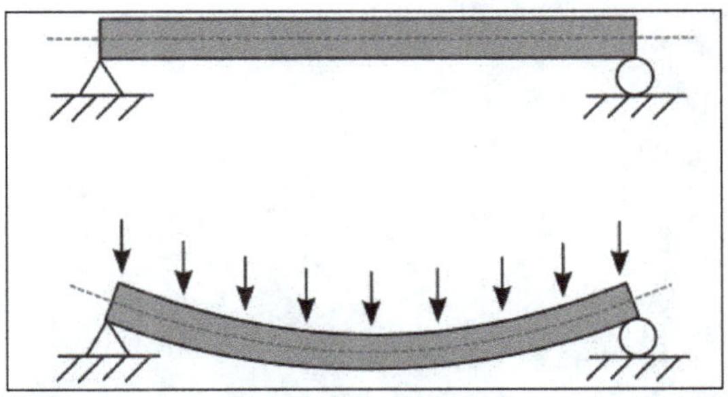

A statically determinate simply supported beam, bending under an evenly distributed load.

Any structure is essentially made up of only a small number of different types of elements:

- Columns
- Beams
- Plates
- Arches
- Shells
- Catenaries

Many of these elements can be classified according to form (straight, plane / curve) and dimensionality (one-dimensional / two-dimensional):

	One-dimensional		**Two-dimensional**	
	straight	**curve**	**plane**	**curve**
(predominantly) bending	beam	continuous arch	plate, concrete slab	lamina, dome
(predominant) tensile stress	rope, tie	Catenary	shell	
(predominant) compression	pier, column		Load-bearing wall	

Columns

Columns are elements that carry only axial force - compression - or both axial force and bending (which is technically called a beam-column but practically, just a column). The design of a column must check the axial capacity of the element, and the buckling capacity.

The buckling capacity is the capacity of the element to withstand the propensity to buckle. Its capacity depends upon its geometry, material, and the effective length of the column, which depends upon the restraint conditions at the top and bottom of the column. The effective length is where is the real length of the column and K is the factor dependent on the restraint conditions.

The capacity of a column to carry axial load depends on the degree of bending it is subjected to, and vice versa. This is represented on an interaction chart and is a complex non-linear relationship.

Beams

Little Belt: a truss bridge in Denmark.

A beam may be defined as an element in which one dimension is much greater than the other two and the applied loads are usually normal to the main axis of the element. Beams and columns are called line elements and are often represented by simple lines in structural modeling.

- Cantilevered (supported at one end only with a fixed connection).
- Simply supported (supported vertically at each end; horizontally on only one to withstand friction, and able to rotate at the supports).
- Fixed (supported at both ends by fixed connection; unable to rotate at the supports).

- Continuous (supported by three or more supports).
- A combination of the above (ex. supported at one end and in the middle).

Beams are elements which carry pure bending only. Bending causes one part of the section of a beam (divided along its length) to go into compression and the other part into tension. The compression part must be designed to resist buckling and crushing, while the tension part must be able to adequately resist the tension.

Trusses

The McDonnell Planetarium by Gyo Obata in St Louis, Missouri, USA, a concrete shell structure.

The 630 foot (192 m) high, stainless-clad (type 304) Gateway Arch in Saint Louis, Missouri.

A truss is a structure comprising members and connection points or nodes. When members are connected at nodes and forces are applied at nodes members can act in tension or in compression. Members acting in compression are referred to as compression members or struts while members acting in tension are referred to as tension members or ties. Most trusses use gusset plates to connect intersecting elements. Gusset plates are relatively flexible and unable to transfer bending moments at the connections, thus allowing the truss members to act primarily in tension or compression.

Trusses are usually utilised in large-span structures, where it would be uneconomical to use solid beams.

Plates

Plates carry bending in two directions. A concrete flat slab is an example of a plate. Plates are understood by using continuum mechanics, but due to the complexity involved they are most often designed using a codified empirical approach, or computer analysis.

They can also be designed with yield line theory, where an assumed collapse mechanism is analysed to give an upper bound on the collapse load. This technique is used in practice but because the method provides an upper-bound, i.e. an unsafe prediction of the collapse load, for poorly conceived collapse mechanisms great care is needed to ensure that the assumed collapse mechanism is realistic.

Shells

Shells derive their strength from their form, and carry forces in compression in two directions. A dome is an example of a shell. They can be designed by making a hanging-chain model, which will act as a catenary in pure tension, and inverting the form to achieve pure compression.

Arches

Arches carry forces in compression in one direction only, which is why it is appropriate to build arches out of masonry. They are designed by ensuring that the line of thrust of the force remains within the depth of the arch. It is mainly used to increase the bountifulness of any structure.

Catenaries

Catenaries derive their strength from their form, and carry transverse forces in pure tension by deflecting (just as a tightrope will sag when someone walks on it). They are almost always cable or fabric structures. A fabric structure acts as a catenary in two directions.

Structural Engineering Theory

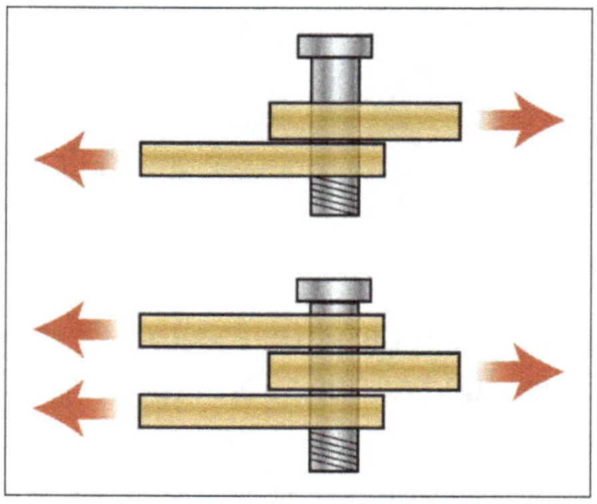

Figure of a bolt in shear stress. Top figure illustrates single shear, bottom figure illustrates double shear.

Structural engineering depends upon a detailed knowledge of applied mechanics, materials science and applied mathematics to understand and predict how structures support and resist self-weight and imposed loads. To apply the knowledge successfully a structural engineer generally requires detailed knowledge of relevant empirical and theoretical design codes, the techniques of structural analysis, as well as some knowledge of the corrosion resistance of the materials and structures, especially when those structures are exposed to the external environment. Since the 1990s, specialist software has become available to aid in the design of structures, with the functionality to assist in the drawing, analyzing and designing of structures with maximum precision; examples include AutoCAD, StaadPro, ETABS, Prokon, Revit Structure, Inducta RCB, etc. Such software may also take into consideration environmental loads, such as from earthquakes and winds.

Materials

Structural engineering depends on the knowledge of materials and their properties, in order to understand how different materials support and resist loads.

Common structural materials are:

- Iron: Wrought iron, Cast iron.
- Concrete: Reinforced concrete, Prestressed concrete.
- Alloy: Steel, Stainless steel.
- Masonry.
- Timber: Hardwood, Softwood.
- Aluminium.
- Composite materials: Plywood.
- Other structural materials:Adobe, Bamboo, Carbon fibre, Fiber reinforced plastic, Mud-brick, Roofing materials.

References

- "ETABS receives "Top Seismic Product of the 20th Century" Award" (PDF). Press Release. Structure Magazine. 2006. Retrieved April 20, 2012
- G. Allen Burton, Jr.; Robert Pitt (2001). Stormwater Effects Handbook: A Toolbox for Watershed Managers, Scientists, and Engineers. New York: CRC/Lewis Publishers. ISBN 0-87371-924-7
- "Architects / Building Engineers in Japan" (PDF). The Japan Architectural Education and Information Center. p. 5. Retrieved 2014-08-16
- O'Flaherty, Coleman A. (2002). Highways: The Location, Design, Construction & Maintenance of Road Pavements. Elsevier. ISBN 0-7506-5090-7
- C. Michael Hogan, "Analysis of highway noise", Journal of Water, Air, & Soil Pollution, Volume 2, Number 3, Biomedical and Life Sciences and Earth and Environmental Science Issue, Pages 387-392, September, 1973, Springer Verlag, Netherlands ISSN 0049-6979
- "Traffic-related Air Pollution near Busy Roads". American Journal of Respiratory and Critical Care Medicine Vol 170. pp. 520-526. 2004
- Ron Bridle; John Porter (2002). The Motorway Achievement: Frontiers of Knowledge and Practice. Thomas Telford. p. 252. ISBN 0-7277-3197-1
- "Some Useful Numbers on the Engineering Properties of Materials (Geologic and Otherwise)" (PDF). Stanford University. Retrieved 2013-12-05

- O'Flaherty, Coleman A. (2002). Highways: The Location, Design, Construction & Maintenance of Road Pavements. Elsevier. p. 252. ISBN 0-7506-5090-7
- Hart-Davis, Adam (2001-06-01). "Roads and surveying". Discovering Roman Technology. BBC.CO.UK. Retrieved 2007-04-22
- Lindeburg, Michael R.; Baradar, Majid (2001). Seismic Design of Building Structures. Professional Publications. ISBN 1-888577-52-5
- "Road Infrastructure Strategic Framework for South Africa". A Discussion Document. National Department of Transport (South Africa). Retrieved 2007-03-24
- Bozorgnia, Yousef; Bertero, Vitelmo V. (2004). Earthquake Engineering: From Engineering Seismology to Performance-Based Engineering. CRC Press. ISBN 978-0-8493-1439-1
- "The Somerset Levels (the oldest timber trackway discovered in Northern Europe)". Current Archaeology 172. Current Archaeology. February 2001. Archived from the original on 2007-03-11. Retrieved 2007-03-25

4

Techniques for Moisture Protection

There are various techniques used in the protection of a building against moisture. This chapter discusses some of them such as building envelope, damp proofing, interstitial condensation, vapor barrier, housewrap, basement waterproofing, etc. The topics elaborated in this chapter will help in gaining a better perspective about moisture protection in construction.

Building Envelope

The building envelope is the physical barrier between the exterior and interior environments enclosing a structure. Generally, the building envelope is comprised of a series of components and systems that protect the interior space from the effects of the environment like precipitation, wind, temperature, humidity, and ultraviolet radiation. The internal environment is comprised of the occupants, furnishings, building materials, lighting, machinery, equipment, and the HVAC (heating, ventilation and air conditioning) system.

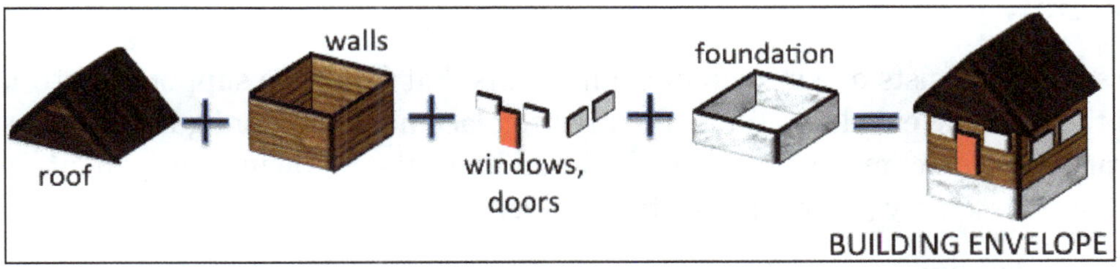

Figure. The components of the building envelope.

Improving the building envelope of houses is one of the best ways to get better energy efficiency.

Function

A building envelope serves many functions. These functions can be divided into 3 categories:

- Support: to ensure strength and rigidity; providing structural support against internal and external loads and forces.
- Control: to control the exchange of water, air, condensation and heat between the interior and exterior of the building.
- Finish: this is for aesthetic purposes. To make the building look attractive while still performing support and control functions.

Physical Components

The building envelope includes the materials that comprise the foundation, wall assembly, roofing systems, glazing, doors, and any other penetrations. The connections and compatibility between these elements is critical to ensure that the building envelope functions as intended.

Foundation

The foundation is the structural component that transmits the loads from the building to the underlying substrate. Typically, some combination of reinforced concrete walls, slabs, and footings constitute the structural components of the foundation. However, the foundation must also be designed to control the transfer of moisture and thermal energy into the interior space.

The transfer of thermal energy through the foundation can be controlled by providing insulation between the interior and exterior environments; however, in some cases the foundation insulation is neglected to reduce construction costs.

Waterproofing the foundation is typically completed by applying a liquid applied asphaltic dampproofing. Additional waterproofing products such as sheet-applied membranes, liquid membranes, cementitious waterproofing, and built-up systems are also viable options.

Drainage around the perimeter of the foundation must be provided to prevent long-term underwater submersion of the waterproofing membrane. One example of a perimeter foundation drain is weeping tile placed in trench complete with gravel ballast backfill, also known as a french drain. In some cases, a sump pit and pump system will be required in addition to the perimeter drain.

Wall Assembly

The wall assembly consists of a system of components that fulfill the support, control, and finish function of the building envelope. While the precise placement and configuration of each component may vary between climates and individual buildings, the following components are typically found in the wall assembly (from exterior to interior):

- Exterior cladding
- Exterior sheathing membrane
- Exterior sheathing
- Insulation
- Structural components
- Vapour barrier
- Interior sheathing

Roofing System

The roofing system is an important part of any house, as it keeps weather out. It consists of shingles on the outside, which are on top of tar sheeting as a vapor barrier. Inside of the tar paper is wood sheathing. Beyond this, the attic areas in most houses are insulated with fiberglass spray insulation. It tends to be fluffy, pink fiberglass. Inhaling fiberglass is extremely bad for a person's respiratory system, so it is important to wear a mask if this insulation type is in one's roofing system.

Glazing

Glazing refers to the panels in windows, doors and skylights - usually glass - that let light through.

Door

Doors are included in the housing envelope, as they tend to be the biggest holes in the envelope. Having outer doors that seal well drastically improves the thermal efficiency of a house.

Other Penetrations

These may include a chimney, or vents for a dryer or stove.

Damp

Structural dampness is the presence of unwanted moisture in the structure of a building, either the result penetration from the outside or condensation within structures. A high proportion of damp problems in buildings are caused by condensation, rain penetration or rising damp.

Breathability and Walls

Old buildings must be allowed to breathe. Whereas modern buildings rely on keeping water out with a system of barriers, buildings that pre-date the mid-19th century are usually constructed of absorbent materials that allow any moisture that enters to evaporate back out.

Because most old buildings were constructed with solid walls without damp proof courses and originally had no roofing felt, rain or below ground moisture could both enter. This did not, however, mean dampness was inevitable. Before central heating was commonplace, heat from open fires drew in large quantities of air through loosely fitting windows and doors. This high rate of ventilation would have quickly evaporated moisture from permeable internal surfaces while the wind dried out any damp roof timbers or permeable external wall surfaces.

An equilibrium was therefore established whereby the moisture being absorbed was equal to that evaporating. When upgrading an old building, you must maintain this equilibrium for the building to work as intended and remain dry.

Causes of Damp

The main risks arise from:

Air Moisture Condensation

Energy-saving measures that reduce ventilation in old buildings such as double-glazing increase relative humidity. Humidity is also raised by modern lifestyles that generate large quantities of water vapour, from bathing, cooking and washing. Condensation will occur on any surface below the dew point (i.e. temperature at which saturated air releases surplus moisture vapour). Interstitial condensation within the pores of materials reduces thermal insulation and further increases the risk of condensation.

Penetrating Damp

Roofs, chimneys, parapets and other exposed parts of a building are most susceptible to rain penetration, especially where access for maintenance is difficult. Junctions in roofs are potential trouble spots, with water exploiting defective lead flashings, mortar fillets, ridges or hips.

Concentrated and prolonged wetting of walls and external joinery arises from poorly maintained rainwater fittings, and leaks from parapet and valley gutters can cause significant damage to structural roof timbers. Hairline cracks in pointing and render invariably admit moisture where cement mortar has been used for repair, rather than lime.

Internal Spillage

This results from overflowing baths or showers, burst pipes, the gradual breakdown of pipe joints, leaks from washing machines or dishwashers, and accidental damage.

Below Ground Damp

This may be rising damp, which is neither as widespread as commonly thought nor a total myth, as sometimes now claimed. Floors become damp where the evaporation of moisture from below is inhibited by vinyl sheet, rubber-backed carpets or other impervious coverings.

New concrete floors or impervious coverings also drive excess moisture into the bases of nearby walls (including chimney stacks), where it rises by capillary action. Damp proof courses were not compulsory in walls prior to 1875 but this is only likely to become a problem where breathability is compromised. In addition to rising damp, below ground moisture can result in problems where ground levels around your building rise unduly.

Tips for Diagnosing Damp

Roofs and Rainwater Fittings

Inspect your roof during wet and windy weather to decide if a damp ceiling patch is due to roof leakage and/or condensation. Debris on the ground (broken slates, tiles and so on) or daylight seen inside lofts indicate possible roof problems.

Defective rainwater fittings may be most obvious during heavy rain, but stains on walls and plant growth provide further clues. Don't forget to check gulleys at ground level.

Condensation is diagnosed from diffuse areas of damp, beads of water droplets on hard shiny surfaces and mould growth on internal finishes. It is intermittent, like penetrating damp, but unrelated to wet weather.

Penetrating damp typically shows up as well-defined patches after heavy rain on south- and west-facing walls. Anticipate moisture ingress through hairline cracks in unsuitable hard, modern cement pointing or rendering.

Below ground moisture causing rising damp can extend up to 900 mm above floor level, sometimes with a classic tidemark on finishes. Salts appear as white deposits but mould growth is rare.

Plumbing

Unusually high water bills (if metered) or a constantly refilling tank may suggest leakage.

Damp Proof Courses and Alternative Systems Compared

Retrospective Damp Proof Courses:

Physical:

- Inserted by cutting in or during rebuilding.
- Can cure rising damp but this drastic method is usually inappropriate.
- Drawbacks: possible major structural problems; potential damage to historic finishes internally; unsuitable for randomly coursed walls; access difficulties; deterioration sometimes of masonry below damp proof course where moisture is concentrated.

Chemical:

- Walls impregnated with chemical solution through holes at bottom to create waterproof barrier.
- Widely used today but not always appropriate for old buildings.
- Drawbacks: drilling holes inadvisable in flint, granite, etc.; hard to form proper barrier in rubble walls with voids; holes unsightly; deterioration sometimes of masonry below damp proof course where moisture concentrated.
- Cost: typically £195/m (including replastering).

Ceramic tubes:

- Holes drilled to receive porous siphons approximately 50mm in diameter that absorb damp and evaporate it from each tube.
- Sound in theory but problems may occur in practice.
- Drawbacks: salt accumulation in tubes may increase moisture; air-flow sometimes inadequate; tubes commonly set in hard cement mortar; unsightly.
- Cost: typically 125/m.

Electro-osmosis:

- Electrical potential aimed at reducing capillary rise using electrodes bedded in wall.
- Cheap but dearth of evidence that electro-osmosis is effective and system rarely used today.
- Drawbacks: adjustment of current needed to match variations in damp along a wall usually impractical.

Other:

- An Austrian product presently under trial in the UK claims to inhibit the passage of water up a wall by inducing a local magnetic field. Achieved non-invasively with unit plugged into mains, typically in loft.
- Likely cost: £3,000/unit (one unit covers an average-sized house).

Investigating Damp

Scientific analysis can be an essential aid for accurately diagnosing a damp problem but the importance of your sight, feel and smell should not be undervalued. Tests include the use of electrical resistance and capacitance meters, on-site moisture testers, hygrometers and salt analysis.

Bear in mind though, that care must be taken when interpreting results. A frequent mistake is to diagnose rising damp purely on the basis of high electrical moisture meter readings. Elevated readings may indicate the presence of salts from past dampness or, if there are no salts, an on-going condensation or possible penetrating damp problem. Continued monitoring is often needed to establish the true cause of a damp problem.

Surveyors have a legal duty to follow a trail of suspicion. Regrettably, many still simply note the occurrence of high meter readings and pass on all responsibility for further investigation to remedial treatment contractors. These contractors have a vested commercial interest, encouraging over-specification. Should a mortgage company insist on work you believe is misguided, challenge this and consider obtaining a written report from an independent surveyor or architect.

Remedial Measures

Effective remedial measures depend on accurate diagnosis, but applying staged remedies can also be part of understanding the cause of a damp problem. Before deciding on more extensive work, the first remedy may involve nothing more than basic maintenance such as clearing a blocked rainwater gulley. Remedies will either cure dampness by addressing the cause (for example, improving drainage) or will manage it by treating the symptoms (changing washing or cooking habits, for instance).

Be sceptical of written guarantees, which are often loaded with get-out clauses and may have no insurance backing. The right approach from your contractor coupled with good workmanship is your best guarantee.

Controlling Air Moisture Condensation

Condensation can be treated by reducing air humidity or keeping surfaces above dew point temperature. Humidity is reduced by cutting the amount of moisture available or increasing ventilation by opening windows, etc. Tumble dryers should be vented to the outside if not of the condenser type, and clothes drying indoors is best avoided.

Temperatures are maintained above dew point with suitable heating. The permanent use of dehumidifiers is a poor substitute for efficient heating and adequate ventilation. Condensation in chimney flues can be eliminated with proper linings. Redundant flues that have been sealed should be fitted with ventilation grilles or re-opened. Lofts should be well insulated and ventilated but make sure insulation does not restrict ventilation at the eaves.

Controlling Penetrating Damp

Repair and Maintain the Roof

Reinstate dislodged and missing slates and tiles before damage occurs to roof timbers or plaster ceilings. SPAB, the Society for the Protection of Ancient Buildings, recommend that renovators avoid spray-on roof foams for the underside of roofs, or external bitumen coatings although other experts disagree. SPABs view is that they prevent proper inspection, hinder the re-use of slates or tiles and, by reducing ventilation, increase the risk of decay.

Brush moss off roofs since it can block gutters and retain moisture, which may damage certain roof coverings in frosty weather. Also, clear gutters and rainwater pipes regularly, particularly if your building is surrounded by trees or perched on by pigeons. Parapet and valley gutters need to be cleared of snow to prevent melt-water rising above them and causing damp internally.

Repair and Protect Walls

Re-point deeply eroded mortar joints in walls. Whilst cement is fine for modern buildings, it is important to use a lime: sand mix (preferably without cement) for most buildings pre-dating about 1900. Localised re-pointing is generally all that is required. Daub, lime mortar or oakum (ships caulking) are useful for closing gaps that may develop around the edges of panel infillings in timber-framed buildings.

Where rain penetrates an exposed south- or west-facing wall, lime wash, lime render and slate or tile hanging are traditional solutions although these cannot be employed without changing the external appearance of the wall. Sometimes installation of a ventilated dry lining system internally is appropriate. The use of colourless water-repellent treatments or plastic-based paints on old masonry is strongly inadvisable.

Controlling Below Ground Damp

The best solution to rising damp may well be to take measures that help your building breathe. Replacing hard cement render or pointing using a more suitable lime-based mortar often improves a damp wall and enables rising damp to dry out. Conversely, the application of waterproof renders and bituminous coatings tends to create or exacerbate damp problems.

Where a floor has a modern damp-proof membrane (horizontal barrier or DPM) that is displacing moisture to the bottoms of walls, it may be sensible to replace this completely with a breathable construction or to at least provide a breathing zone for evaporation around the perimeter of the room. When underfloor heating is being installed, there are many situations where DPM-insertion can be avoided by employing materials such as lime concrete and expanded clay insulation.

Reducing or removing the source of moisture may also help alleviate rising damp. French drains can be an effective and relatively inexpensive answer but it is preferable not to site them directly against walls and rodding points must be provided. Otherwise, blockages can effectively convert them into a sump and increase dampness. Consider also the structural and archaeological implications.

Although retrofit DPCs can sometimes be appropriate, with an old building always consider first whether rising damp is actually too minor to matter and, if it is significant, whether more sympathetic ways exist of dealing with it. Where any timber is at risk of decay, for example, you might be able to simply isolate it. Similarly, the eradication of any contributing moisture from other sources such as rainsplash off closely abutting patios could obviate the need for more extensive remedial treatment.

Damp can be particularly troublesome in cellars but increased ventilation (including opening up redundant flues), re-pointing and lowering the water table locally can be effective. Failing this, it may be worth considering a dry lining system. Tanking (applying waterproof linings to walls and floors) is not recommended in old buildings.

Internal Finishes

To minimise the risk of future problems, lime plaster should usually be used for any replastering rather than the anti-sulphate or renovating plasters favoured by many treatment companies. Decoration with paints such as limewash and soft distemper, where possible, will maximise breathability. Old items of joinery removed during work should be carefully repaired and reinstated, not automatically replaced.

Damp Proofing

Damp proofing is a general term that covers methods and treatments used to prevent damp from being absorbed through walls or floors into the interior of a property.

Any property can be subject to damp problems, especially older properties which may have been constructed without a damp-proof membrane. Whether it's rising damp or penetrating damp, the property care specialists are experts at identifying the types of damp within a property as well as potential problems.

Types of Damp Protection

The two types of protection methods for damp proofing residential and commercial properties are Damp Proof Course (DPC) and Damp Proof Membrane (DPM).

Most properties we live in today should have evidence of a damp proof course about 150mm (6inch) above ground level. This may be seen as a slightly thicker mortar course with a slate or bitumen sheet poking through. These damp course materials are fairly resilient, unless movement of the building subsidence causes them to crack.

There is sometimes confusion between what a damp proof membrane is and what a damp proof course is, particularly as they can be used together. We advise linking the two to allow a building to be fully protected from the damp, wet ground around it.

Damp Proof Course

A damp proof course (DPC) is one of several damp proofing treatments used to prevent damp problems developing within a property. Damp proof course repair can be applied using a variety of different methods and is a long term solution to preclude moisture from entering a property through walls. The build-up of excess moisture within a building can eventually result in structural damage and therefore pose a risk to your property.

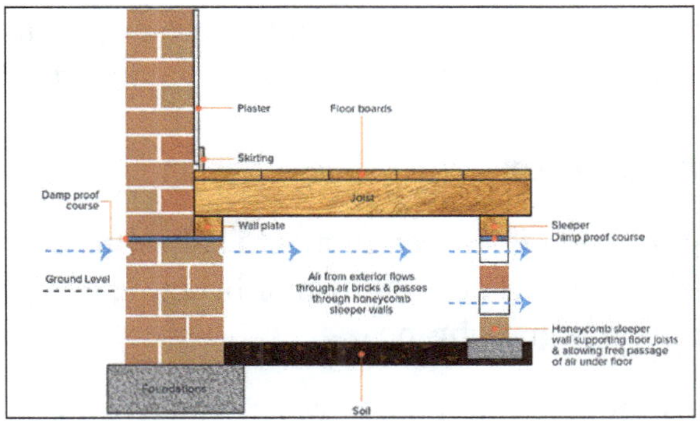

The property care specialists are experts at identifying the types of damp within a property as well as potential problems and will apply the most suitable treatment based on damp proof course regulations. All our injected damp proof course treatments come with a 30 year damp proof course guarantee to ensure a safe and stable property.

Damp Proof Membrane

A damp proof membrane (DPM) is another common method used to prevent rising damp from occurring within a property. The damp proof membrane sheets are made from materials such as polyethylene or butyl rubber and act as a successful barrier to prevent damp from making it's way into a property. The well-qualified technicians have a great understanding in how to resolve damp issues and are experienced in installing damp proof membranes for walls and floors within a property.

Working of Damp Proof Membrane

A sheet of material (impervious to water) laid in one piece that is placed beneath the concrete floor of a property to prevent groundwater seeping upwards through the concrete base. A common example is polyethylene sheeting laid under a concrete slab.

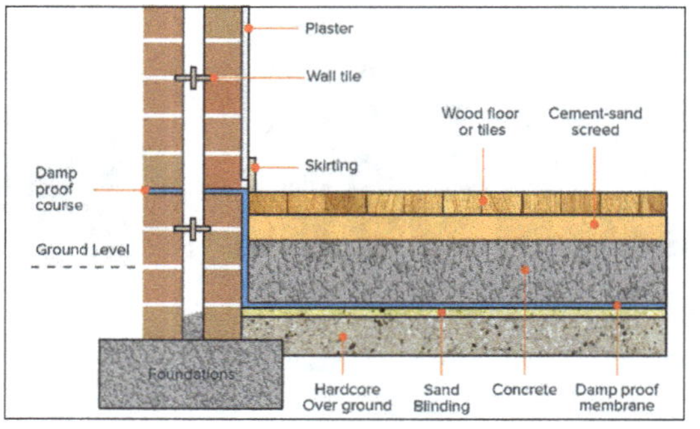

Damp Proofing Walls

The build-up of excess moisture within a building can lead to damp walls which can eventually result into a handful of problems. If you notice signs of damp on walls within your property then we strongly advise you act fast to resolve the issue. Damp on internal walls is unsightly, unhealthy and is the primary cause of wood decay in a building, which can put your property at risk of structural damage.

The property care specialists have many years' experience in treating damp walls and are experts at applying the appropriate damp proof course based on your requirements. Damp proofing internal walls come as a second nature to us so you can be sure your property is in safe hands.

Damp Proofing Concrete Floor

We offer professional damp concrete floor treatments to help eliminate rising damp from within a property. This can be caused by the lack, breakdown or bridging of a physical damp course, or the absence or damage of a damp proof membrane.

The damp proof membranes for concrete floors are an effective solution to counter rising damp and at Rentokil, the property care specialists are registered TrustMark contractors and proud members of the PCA so you know that your property is in safe hands.

Remedial Damp Protection

Issues with your current damp protection can occur due to subsidence or the result of long term deterioration which allows damp to rise through the walls. If you begin to notice:

- Peeling paint or wallpaper
- Discolouration
- Decaying skirting boards and timber floors
- Salt staining
- Crumbling plaster

The Certificated Surveyor in Remedial Treatment can conduct a professional damp survey to identify the cause of damp. Rentokil's remedial damp proof course can then be installed to ensure your property is damp protected.

Interstitial Condensation

Interstitial condensation is very simply, condensation within the structure of a building, i.e. not on the surface and not clearly visible until often, it's too late and the damage is done. Warm air can hold a lot of moisture, so air within a dwelling for example, is perfect for moisture-laden air, especially as we create so much of it – through washing, cooking, breathing, etc. Condensation occurs where this warm, moisture-laden air is suddenly cooled and can no longer hold the high levels of moisture. In a building, this is usually when the warm air meets a cold surface such as a mirror, window or external wall. If the wall is uninsulated, then it can be extremely cold in winter time and condensation will be visible on the surface.

But what happens if the wall is fitted with internal wall insulation (dry lining), for example? The moisture contained in air is water vapour and is usually completely invisible to the human eye, as it is in such tiny particles. This water vapour can easily pass through most building materials such as plaster, slabs, insulation, etc. And so, they do just that.

So, Our wall is insulated on the inside. This makes the external wall even colder, as now the heat cannot get out to heat it up. But the moisture can. Now, the moisture-laden air is cooled even further than before and dumps even more condensation, down along the wall surface, exactly as before, only now, the old wall surface is within the overall structure of the wall (including the dry-lining).

This is called "interstitial" condensation, as it happens within the wall's structure. Any moisture within a structure is not good. It leads, first of all, to nasty moulds and fungi to grow within the structure which leads to very poor air quality within the building. This can have severe health implications. Dampness also leads to structural damage, whether it's rotting timber, degraded insulation, cables or pipes/fitting corroding, etc. It's a huge problem because it can't be seen for maybe several years but by then, it's far too late to do anything about it. This can equally be a problem in new and existing buildings, so great care should be taken, especially when altering the make-up of any building element e.g. wall.

The good news is that correct construction methods such as: using an appropriate "vapour check", a barrier against water vapour; placing the insulation in a different position, such as outside the wall, or introducing the appropriate ventilation where condensation may occur; can eliminate the possibility of interstitial condensation. It is critically important to consider what construction methods are employed before work is carried out. Unfortunately, unscrupulous or ignorant contractors may try to lead we down the wrong path as the incorrect and easy methods are generally much cheaper, quicker and easier to install.

Vapor Barrier

The function of a vapor barrier is to retard the migration of water vapor. Where it is located in an assembly and its permeability is a function of climate, the characteristics of the materials that comprise the assembly and the interior conditions. Vapor barriers are not typically intended to retard the migration of air. That is the function of air barriers.

Confusion on the issue of vapor barriers and air barriers is common. The confusion arises because air often holds a great deal of moisture in the vapor form. When this air moves from location to location due to an air pressure difference, the vapor moves with it. This is a type of migration of water vapor. In the strictest sense air barriers are also vapor barriers when they control the transport of moisture-laden air.

Vapor barriers are also a cold climate artifact that have diffused into other climates more from ignorance than need. The history of cold climate vapor barriers itself is a story based more on personalities than physics. Rose regales readers of this history. It is frightening indeed that construction practices can be so dramatically influenced by so little research and reassuring indeed that the inherent robustness of most building assemblies has been able to tolerate such foolishness.

Problems

Incorrect use of vapor barriers is leading to an increase in moisture related problems. Vapor barriers were originally intended to prevent assemblies from getting wet. However, they often prevent assemblies from drying. Vapor barriers installed on the interior of assemblies prevent assemblies from drying inward. This can be a problem in any air-conditioned enclosure. This can be a problem in any below grade space. This can be a problem when there is also a vapor barrier on the exterior. This can be a problem where brick is installed over building paper and vapor permeable sheathing.

Measures Need to be Taken

Two seemingly simple requirements for building enclosures bedevil engineers and architects almost endlessly:

- Keep water out
- Let water out if it gets in

Water can come in several phases: liquid, solid, vapor and adsorbed. The liquid phase as rain and ground water has driven everyone crazy for hundreds of years but can be readily understood - drain everything and remember the humble flashing. The solid phase also drives everyone crazy when we have to shovel it or melt it, but at least most professionals understand the related building problems (ice damming, frost heave, freeze-thaw damage). But the vapor phase is in a class of craziness all by itself. We will conveniently ignore the adsorbed phase and leave it for someone else to deal with. Note that adsorbed water is different than absorbed water.

The fundamental principle of control of water in the liquid form is to drain it out if it gets in – and let us make it perfectly clear – it will get in if you build where it rains or if you put your building in the ground where there is water in the ground. This is easy to understand, logical, with a long historical basis.

The fundamental principle of control of water in the solid form is to not let it get solid and if it does – give it space or if it is solid not let it get liquid and if it does drain it away before it can get solid again. This is a little more difficult to understand, but logical and based on solid research. Examples of this principle include the use of air entrained concrete to control freeze-thaw damage and the use of attic venting to provide cold roof decks to control ice damming.

The fundamental principle of control of water in the vapor form is to keep it out and to let it out if it gets in. Simple right? No chance. It gets complicated because sometimes the best strategies to keep water vapor out also trap water vapor in. This can be a real problem if the assemblies start out wet because of rain or the use of wet materials.

It gets even more complicated because of climate. In general water vapor moves from the warm side of building assemblies to the cold side of building assemblies. This is simple to understand, except we have trouble deciding what side of a wall is the cold or warm side. Logically, this means we need different strategies for different climates. We also have to take into account differences between summer and winter.

Finally, complications arise when materials can store water. This can be both good and bad. A cladding system such as a brick veneer can act as a reservoir after a rainstorm and significantly complicate wall design. Alternatively, wood framing or masonry can act as a hygric buffer absorbing water lessening moisture shocks.

What is required is to define vapor control measures on a more regional climatic basis and to define the vapor control measures more precisely.

Part of the problem is that we struggle with names and terms. We have vapor retarders, we have vapor barriers, we have vapor permeable we have vapor impermeable, etc. What do these terms mean? It depends on whom you ask and whether they are selling something or arguing with a building official. In an attempt to clear up some of the confusion the following definitions are proposed:

Vapor Retarder: The element that is designed and installed in an assembly to retard the movement of water by vapor diffusion.

The unit of measurement typically used in characterizing the water vapor permeance of materials is the "perm." It is further proposed here that there should be several classes of vapor retarders (this is nothing new – it is an extension and modification of the Canadian General Standards Board approach that specifies Type I and Type II vapor retarders – the numbers here are a little different however):

Class I Vapor Retarder:	0.1 perm or less
Class II Vapor Retarder:	1.0 perm or less and greater than 0.1 perm
Class III Vapor Retarder	10 perm or less and greater than 1.0 perm
Test Procedure for vapor retarders:	ASTM E-96 Test Method A (the desiccant method or dry cup method)

Finally, a vapor barrier is defined as:

Vapor Barrier: A Class I vapor retarder.

The current International Building Code (and its derivative codes) defines a vapor retarder as 1.0 perm or less (using the same test procedure). In other words the current code definition of a vapor retarder is equivalent to the definition of a Class II Vapor Retarder proposed by the author.

Continuing in the spirit of finally defining terms that are tossed around in the enclosure business. It is also proposed that materials be separated into four general classes based on their permeance:

Vapor impermeable	0.1 perm or less
Vapor semi-impermeable	1.0 perm or less and greater than 0.1 perm
Vapor semi-permeable:	10 perms or less and greater than 1.0 perm
Vapor permeable:	greater than 10 perms

Recommendations for Building Enclosures

The following building assembly recommendations are climatically based and are sensitive to cladding type (brick or stone veneer, stucco) and structure (concrete block, steel or wood frame, precast concrete).

The recommendations apply to residential, business, assembly, and educational and mercantile occupancies. The recommendations do not apply to special use enclosures such as spas, pool buildings, museums, hospitals, data processing centers or other engineered enclosures such as factory, storage or utility enclosures.

The recommendations are based on the following principles:

- Avoidance of using vapor barriers where vapor retarders will provide satisfactory performance. Avoidance of using vapor retarders where vapor permeable materials will provide satisfactory performance. Thereby encouraging drying mechanisms over wetting prevention mechanisms.
- Avoidance of the installation of vapor barriers on both sides of assemblies – i.e. "double vapor barriers" in order to facilitate assembly drying in at least one direction.
- Avoidance of the installation of vapor barriers such as polyethylene vapor barriers, foil faced batt insulation and reflective radiant barrier foil insulation on the interior of air-conditioned assemblies – a practice that has been linked with moldy buildings.
- Avoidance of the installation of vinyl wall coverings on the inside of air-conditioned assemblies – a practice that has been linked with moldy buildings.
- Enclosures are ventilated meeting ASHRAE Standard 62.2 or 62.1.

Each of the recommended building assemblies were evaluated using dynamic hygrothermal modeling. The moisture content of building materials that comprise the building assemblies all remained below the equilibrium moisture content of the materials as specified in ASHRAE 160 P under this evaluation approach. Interior air conditions and exterior air conditions as specified by ASHRAE 160 P were used. WUFI was used as the modeling program.

More significantly, each of the recommended building assemblies have been found by the author to provide satisfactory performance under the limitations noted. Satisfactory performance is defined as no moisture problems reported or observed over at least a 10-year period.

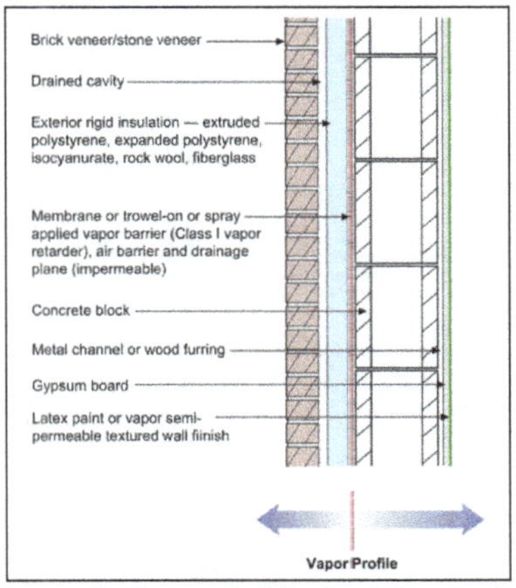

Figure A: Concrete block with exterior insulation and brick or stone veneer.

Applicability – All hygro-thermal regions.

This is arguably the most durable wall assembly available to architects and engineers. It is constructed from non-water sensitive materials and due to the block construction has a large moisture storage (or hygric buffer) capacity. It can be constructed virtually anywhere. In cold climates condensation is limited on the interior side of the vapor barrier as a result of installing all of the thermal insulation on the exterior side of the vapor barrier (which is also the drainage plane and air barrier in this assembly). In hot climates any moisture that condenses on the exterior side of the vapor barrier will be drained to the exterior since the vapor barrier is also a drainage plane. This wall assembly will dry from the vapor barrier inwards and will dry from the vapor barrier outwards.

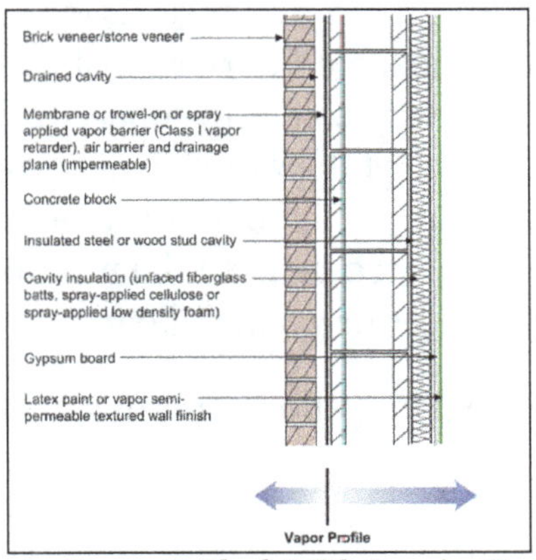

Figure B: Concrete Block With Interior Frame Wall Cavity Insulation and Brick or Stone Veneer.

Applicability – Limited to mixed-humid, hot-humid, mixed-dry, hot-dry and marine regions – should not be used in cold, very cold, and subarctic/arctic regions.

This wall assembly has all of the thermal insulation installed to the interior of the vapor barrier and therefore should not be used in cold regions or colder. It is also a durable assembly due to the block construction and the associated moisture storage (hygric buffer) capacity. The wall assembly does contain water sensitive cavity insulation (except where spray foam is used) and it is important that this assembly can dry inwards – therefore vapor semi impermeable interior finishes such as vinyl wall coverings should be avoided. In this wall assembly the vapor barrier is also the drainage plane and air barrier.

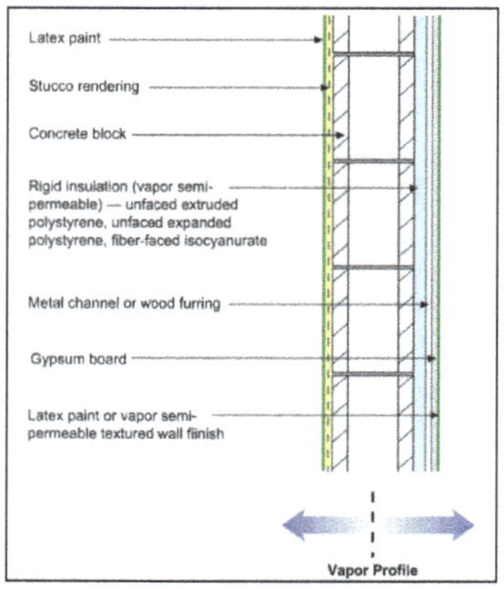

Figure C: Concrete block with interior rigid insulation and stucco.

Applicability – All hygro-thermal regions*

This assembly has all of the thermal insulation installed on the interior of the concrete block construction but differs from figure B since it does not have a vapor barrier on the exterior. The assembly also does not have a vapor barrier on the interior of the assembly. It has a large moisture storage (hygric buffer) capacity due to the block construction. The rigid insulation installed on the interior should ideally be non-moisture sensitive and allow the wall to dry inwards – hence the recommended use of vapor semi permeable foam sheathing. Note that foam sheathing faced with aluminum foil or polypropylene skins would also be acceptable provided only non-moisture sensitive materials are used at the masonry block to insulation interface. It is important that this assembly inboard of the foam sheathing can dry inwards except in very cold and subarctic/arctic regions – therefore vapor semi impermeable interior finishes such as vinyl wall coverings should be avoided in assemblies – except in very cold and subarctic/arctic regions. Vapor impermeable foam sheathings should be used in place of the vapor semi permeable foam sheathings in very cold and subarctic/arctic regions. The drainage plane in this assembly is the latex painted stucco rendering. A Class III vapor retarder is located on both the interior and exterior of the assembly (the latex paint on the stucco and on the interior gypsum board).

* In very cold and subarctic/arctic regions vapor impermeable foam sheathings are recommended

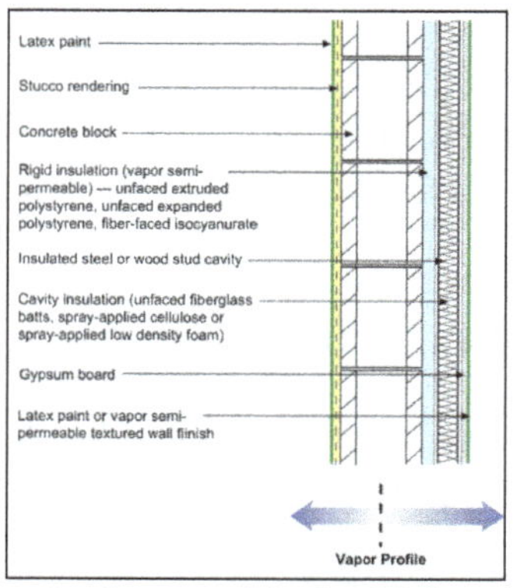

Figure D: Concrete block with interior rigid insulation/frame wall with cavity insulation and stucco.

Applicability – All hygro-thermal regions*

This assembly is a variation of figure C. It also has all of the thermal insulation installed on the interior of the concrete block construction but differs from figure C, due to the addition of a frame wall to the interior of the rigid insulation. This assembly also does not have a vapor barrier on the exterior. The assembly also does not have a vapor barrier on the interior of the assembly. It has a large moisture storage (hygric buffer) capacity due to the block construction. The rigid insulation installed on the interior should ideally be non-moisture sensitive and allow the wall to dry inwards — hence the recommended use of vapor semi permeable foam sheathing. Note that foam sheathing faced with aluminum foil or polypropylene skins would also be acceptable provided only non-moisture sensitive materials are used at the masonry block to insulation interface. It is important that this assembly inboard of the rigid insulation can dry inwards even in very cold and subarctic/arctic regions — therefore vapor semi impermeable interior finished such as vinyl wall coverings should be avoided in assemblies. Vapor impermeable foam sheathings should be used in place of the vapor semi permeable foam sheathings in very cold and subarctic/arctic regions. The drainage plane in this assembly is the latex painted stucco rendering. A Class III vapor retarder is located on both the interior and exterior of the assembly (the latex paint on the stucco and on the interior gypsum board).

* In very cold and sub arctic/arctic regions vapor impermeable foam sheathings are recommended – additionally the thickness of the foam sheathing should be determined by hygro-thermal analysis so that the interior surface of the foam sheathing remains above the dew point temperature of the interior air.

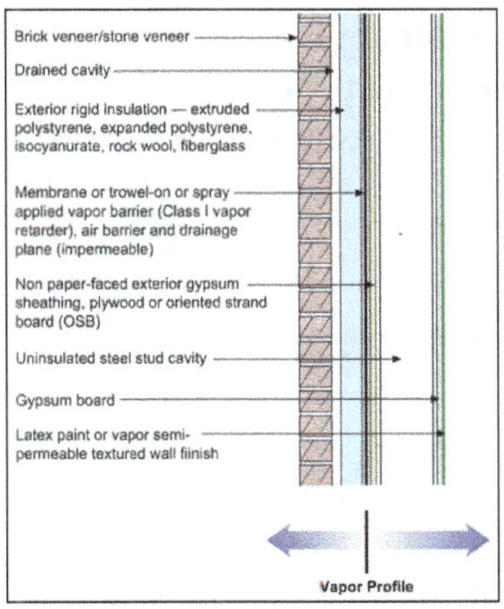

Figure E: Frame wall with exterior insulation and brick of stone veneer.

Applicability – All hygro-thermal regions

This wall is a variation of figure A, but without the moisture storage (or hygric buffer) capacity. This wall is also a durable wall assembly. It is constructed from non-water sensitive materials and has a high drying potential inwards due to the frame wall cavity not being insulated. It can also be constructed virtually anywhere. In cold climates condensation is limited on the interior side of the vapor barrier as a result of installing all of the thermal insulation on the exterior side of the vapor barrier (which is also the drainage plane and air barrier in this assembly). In hot climates any moisture that condenses on the exterior side of the vapor barrier will be drained to the exterior since the vapor barrier is also a drainage plane. This wall assembly will dry from the vapor barrier inwards and will dry from the vapor barrier outwards.

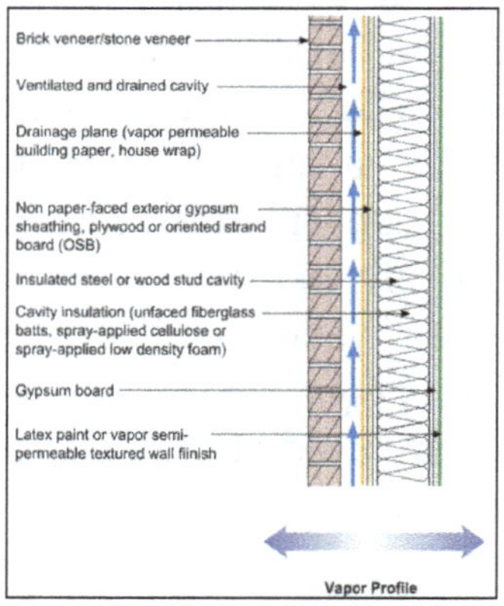

Figure F: Frame Wall With Cavity Insulation and Brick or Stone Veneer.

Applicability – Limited to mixed-humid, hot-humid, mixed-dry, hot-dry and marine regions – can be used with hygro-thermal analysis in some areas in cold regions- should not be used in very cold and subarctic/arctic regions

This wall is a flow through assembly – it can dry to both the exterior and the interior. It has a Class III vapor retarder on the interior of the assembly (the latex paint on the gypsum board). It is critical in this wall assembly that the exterior brick veneer (a "reservoir" cladding) be uncoupled from the wall assembly with a ventilated and drained cavity. The cavity behind the brick veneer should be at least 2 inches wide and free from mortar droppings. It must also have air inlets ("weep holes") at its base and air outlets ("weep holes") at its top in order to provide back ventilation of the brick veneer. The drainage plane in this assembly is the building paper or building wrap. The air barrier can be any of the following: the interior gypsum board, the exterior gypsum wallboard or the exterior building wrap.

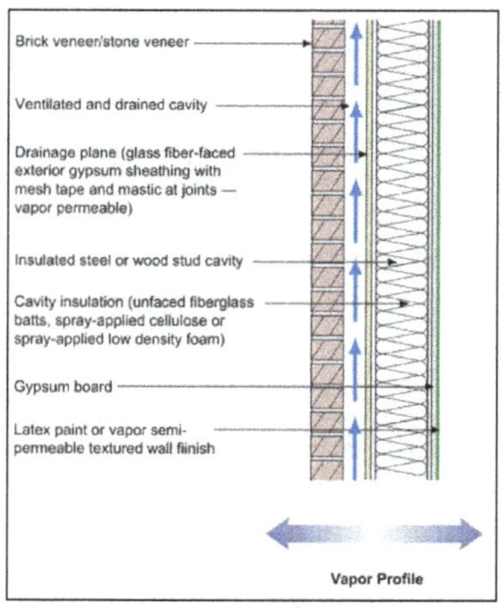

Figure G: Frame Wall With Cavity Insulation and Brick or Stone Veneer.

Applicability – Limited to mixed-humid, hot-humid, mixed-dry, hot-dry and marine regions – can be used with hygro-thermal analysis in some areas in cold regions - should not be used in very cold and subarctic/arctic regions

This wall is a variation of figure F. The exterior gypsum sheathing becomes the drainage plane. As in figure F, this wall is a flow through assembly – it can dry to both the exterior and the interior. It has a Class III vapor retarder on the interior of the assembly (the latex paint on the gypsum board). It is also critical in this wall assembly that the exterior brick veneer (a "reservoir" cladding) be uncoupled from the wall assembly with a ventilated and drained cavity. The cavity behind the brick veneer should be at least 2 inches wide and free from mortar droppings. It must also have air inlets ("weep holes") at its base and air outlets ("weep holes") at its top in order to provide back ventilation of the brick veneer. The air barrier in this assembly can be either the interior gypsum board or the exterior gypsum sheathing.

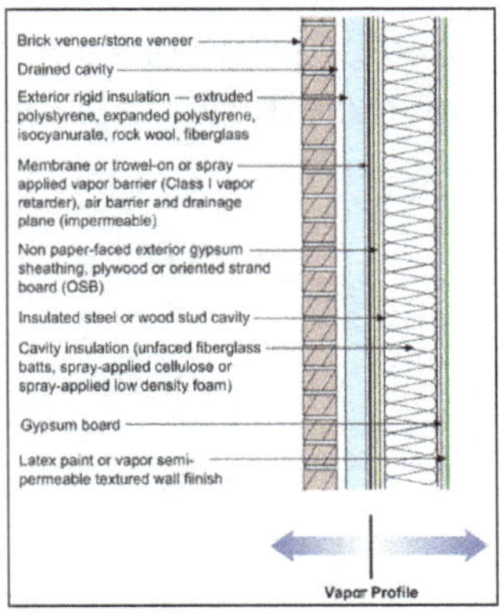

Figure H: Frame Wall With Exterior Rigid InsulationWith Cavity Insulation and Brick or Stone Veneer.

Applicability – All hygro-thermal regions except subarctic/arctic – in cold and very cold regions the thickness of the foam sheathing should be determined by hygro-thermal analysis so that the interior surface of the foam sheathing remains above the dew point temperature of the interior air.

This wall is a variation of figure E. In cold climates condensation is limited on the interior side of the vapor barrier as a result of installing some of the thermal insulation on the exterior side of the vapor barrier (which is also the drainage plane and air barrier in this assembly). In hot climates any moisture that condenses on the exterior side of the vapor barrier will be drained to the exterior since the vapor barrier is also a drainage plane. This wall assembly will dry from the vapor barrier inwards and will dry from the vapor barrier outwards. Since this wall assembly has a vapor barrier that is also a drainage plane it is not necessary to back vent the brick veneer reservoir cladding as in figures F and figure G. Moisture driven inwards out of the brick veneer will condense on the vapor barrier/drainage plane and be drained outwards.

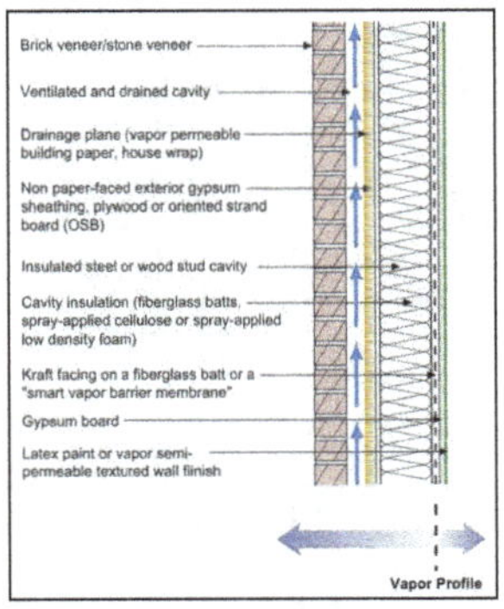

Figure I: Frame Wall With Cavity Insulation and Brick or Stone Veneer With Interior Vapor Retarder.

Applicability – Limited to cold and very cold regions.

This wall is a variation of figure F except it has a Class II vapor retarder on the interior limiting its inward drying potential – but not eliminating it. It still considered a flow through assembly – it can dry to both the exterior and the interior. It is critical in this wall assembly- as figure F and figure G- that the exterior brick veneer (a “reservoir” cladding) be uncoupled from the wall assembly with a ventilated and drained cavity. The cavity behind the brick veneer should be at least 2 inches wide and free from mortar droppings. It must also have air inlets (“weep holes”) at its base and air outlets (“weep holes”) at its top in order to provide back ventilation of the brick veneer. The drainage plane in this assembly is the building paper or building wrap. The air barrier can be any of the following: the interior gypsum board, the exterior gypsum board or the exterior building wrap.

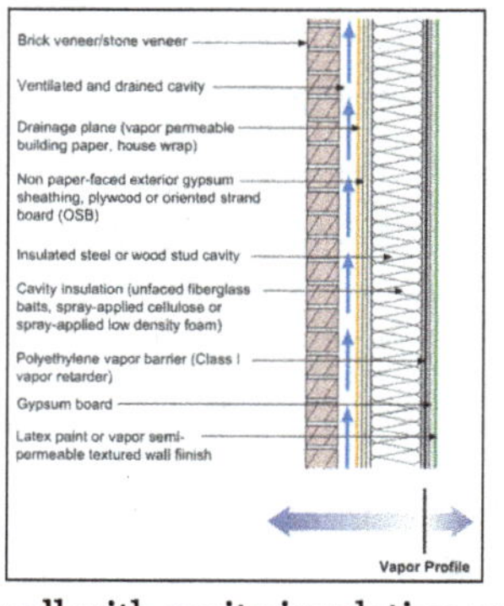

Figure J: Frame wall with cavity insulation and brick or stone veneer with interior vapor barrier.

This wall is a further variation of figure F, but now it has a Class I vapor retarder on the interior (a “vapor barrier”) completely eliminating any inward drying potential. It is considered the “classic” cold climate wall assembly as in figure F, figure G and figure H. It is critical in this wall assembly that the exterior brick veneer (a “reservoir” cladding) be uncoupled from the wall assembly with a ventilated and drained cavity. The cavity behind the brick veneer should be at least 2 inches wide and free from mortar droppings. It must also have air inlets at its base and air outlets at its top in order to provide back ventilation of the brick veneer. The drainage plane in this assembly is the building paper or building wrap. The air barrier can be any of the following: the interior polyethylene vapor barrier, the interior gypsum board, the exterior gypsum board or the exterior building wrap.

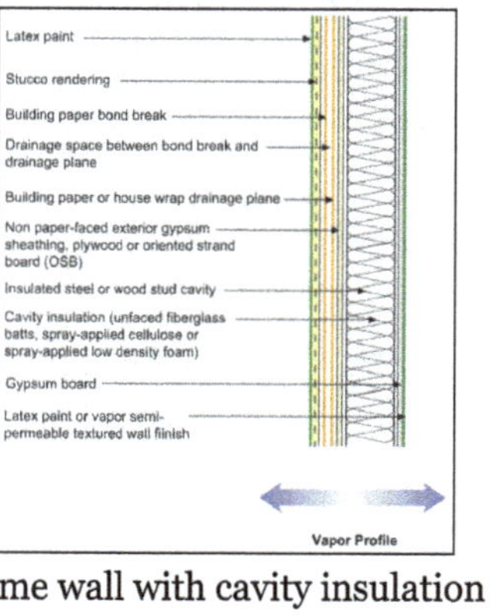

Figure K: Frame wall with cavity insulation and stucco.

Applicability – Limited to mixed-humid, hot-humid, mixed-dry, and hot-dry regions should not be used in marine, cold, very cold, and subarctic/arctic regions.

This wall is also a flow through assembly similar to figure F, but without the brick veneer – it has a stucco cladding. It can dry to both the exterior and the interior. It has a Class III vapor retarder on the interior of the assembly (the latex paint on the gypsum board). It is critical in this wall assembly that a drainage space be provided between the stucco rendering and the drainage plane. This can be accomplished by installing a bond break (a layer of tar paper) between the drainage plane and the stucco. A spacer mat can also be used to increase drainability. Alternatively, a textured or profiled drainage plane (building wrap) can be used. The drainage plane in this assembly is the building paper or building wrap. The air barrier can be any of the following: the interior gypsum board, the exterior stucco rendering, the exterior sheathing or the exterior building wrap.

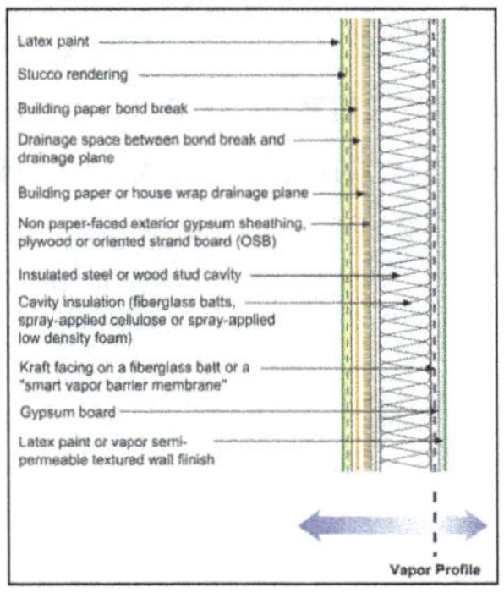

Figure L: Frame Wall With Cavity Insulation and Stucco With Interior Vapor Retarder.

Applicability – Limited to marine, cold and very cold regions

This wall is a variation of figure F and figure K, except it has a Class II vapor retarder on the interior limiting its inward drying potential – but not eliminating it. It still considered a flow through assembly – it can dry to both the exterior and the interior. It is critical in this wall assembly – as in Figure K – that a drainage space be provided between the stucco rendering and the drainage plane. This can be accomplished by installing a bond break (a layer of tar paper) between the drainage plane and the stucco. A spacer mat can also be used to increase drainability. Alternatively, a textured or profiled drainage plane (building wrap) can be used. The drainage plane in this assembly is the building paper or building wrap. The air barrier can be any of the following: the interior gypsum board, the exterior stucco rendering, the exterior sheathing or the exterior building wrap.

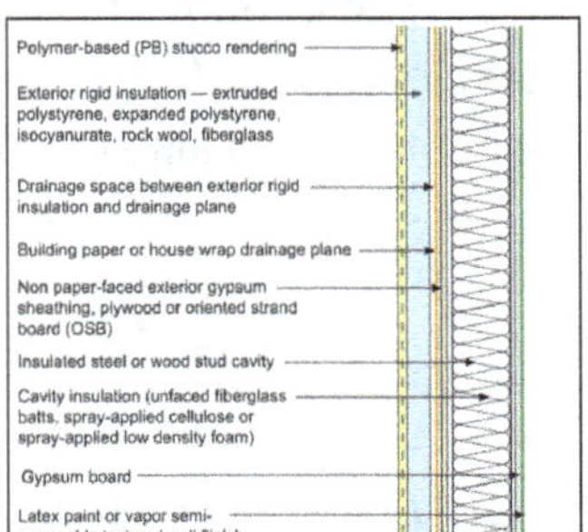

Figure M: Frame wall with exterior rigid insulation with cavity insulation and stucco.

Applicability – All hygro-thermal regions except subarctic/arctic – in cold and very cold regions the thickness of the foam sheathing should be determined by hygro-thermal analysis so that the interior surface of the foam sheathing remains above the dew point temperature of the interior air.

This is a water managed exterior insulation finish system (EIFS). Unlike "face-sealed" EIFS this wall has a drainage plane inboard of the exterior stucco skin that is drained to the exterior. It is also a flow through assembly similar to Figure F. It can dry to both the exterior and the interior. It has a Class III vapor retarder on the interior of the assembly (the latex paint on the gypsum board). It is critical in this wall assembly that a drainage space be provided between the exterior rigid insulation and the drainage plane. This can be accomplished by installing a spacer mat or by providing drainage channels in the back of the rigid insulation. Alternatively, a textured or profiled drainage plane (building wrap) can be used. The drainage plane in this assembly is the building paper or building wrap. The air barrier can be any of the following: the interior gypsum board, the exterior stucco rendering, the exterior sheathing or the exterior building wrap.

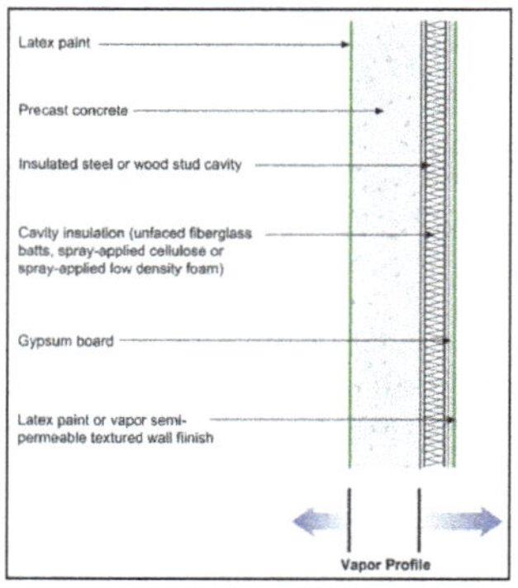

Figure N: Precast concrete with interior frame wall cavity insulatioin.

*Applicability – Limited to mixed-humid, hot-humid, mixed-dry, hot-dry and marine regions – should not be used in cold, very cold, and subarctic/arctic regions.

The vapor barrier in this assembly is the precast concrete itself. Therefore this wall assembly has all of the thermal insulation installed to the interior of the vapor barrier. Of particular concern is the fact that the thermal insulation is air permeable (except where spray foam is used). Therefore this wall assembly should not be used in cold regions or colder. It has a small moisture storage (hygric buffer) capacity due to the precast concrete construction. The wall assembly does contain water sensitive cavity insulation (except where spray foam is used) and it is important that this assembly can dry inwards – therefore vapor semi impermeable interior finishes such as vinyl wall coverings should be avoided. In this wall assembly the precast concrete is also the drainage plane and air barrier.

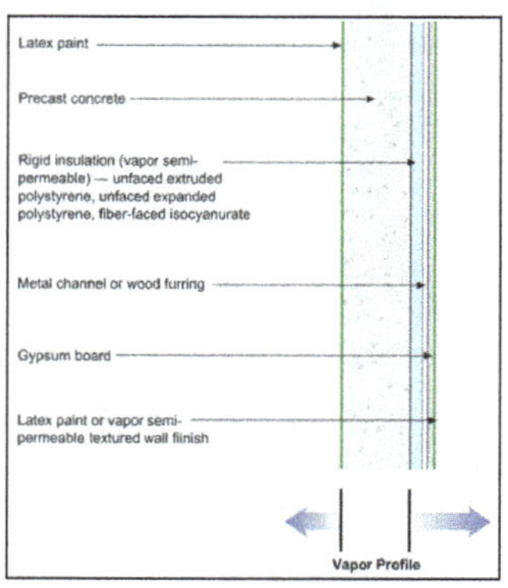

Figure O: Precast Concrete With Interior Rigid Insulation.

This assembly has all of the thermal insulation installed on the interior of the precast concrete. The assembly also does not have a vapor barrier on the interior of the assembly. It has a small moistuure storage (hygric buffer) capacity due to the precast concrete construction. The rigid insulation installed on the interior should ideally be non-moisture sensitive and allow the wall to dry inwards — hence the recommended use of vapor semi permeable foam sheathing. Note that foam sheathing faced with aluminum foil or polypropylene skins would also be acceptable provided only non-moisture sensitive materials are used at the concrete to insulation interface. It is important that this assembly inboard of the foam sheathing can dry inwards except in very cold subarctic/arctic regions — therefore vapor semi impermeable interior finishes such as vinyl wall coverings should be avoided in assemblies — except in very cold and subarctic/arctic regions. Vapor impermeable foam sheathings should be used in place of the vapor semi permeable foam sheathings in very cold and subarctic/arctic regions. The drainage plane in this assembly is the latex painted precast concrete. A Class III vapor retarder is located on both the interior and exterior of the assembly (the latex paint on the stucco and on the interior gypsum board).

* In very cold and subarctic/arctic regions vapor impermeable foam sheathings are recommended

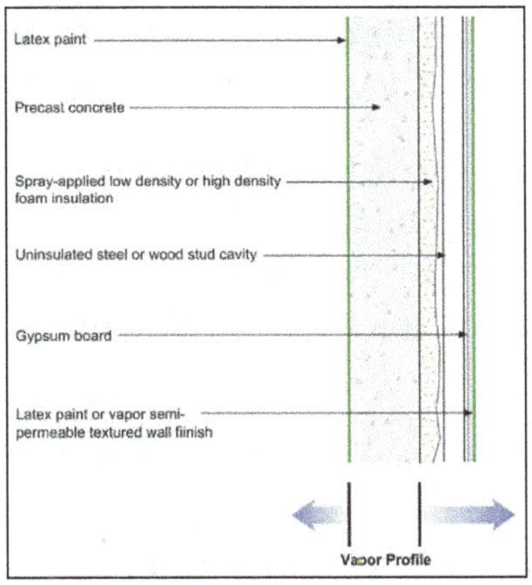

Figure P: Precast Concrete With Interior Spray Applied Foam Insulation.

Applicability – All hygro-thermal regions*

This assembly has all of the thermal insulation installed on the interior of the precast concrete. The assembly also does not have a vapor barrier on the interior of the assembly. It has a small moisture storage (hygric buffer) capacity due to the precast concrete construction. The spray foam insulation installed on the interior of the precast concrete is non-moisture sensitive and allows the wall to dry inwards. It is important that this assembly can dry inwards except in very cold and subarctic/arctic regions – therefore vapor semi impermeable interior finishes such as vinyl wall coverings should be avoided in assemblies – except in very cold and subarctic/arctic regions. High-density spray foam, due to its vapor semi impermeable characteristics should be used in place of low-density foam in very cold and subarctic/arctic regions. The drainage plane in this assembly is the latex painted precast concrete. A Class III vapor retarder is located on both the interior and exterior of the assembly (the latex paint on the stucco and on the interior gypsum board.

* In very cold and subarctic/arctic regions high-density spray foam (vapor semi impermeable) is recommended.

Housewrap

Contractors use house wrap to cover the wood frame before the application of stone, wood, brick, or vinyl siding. House wrap helps keep moisture away from the interior of the house and protects the surface of the exterior walls.

House wrap is made with water-resistant materials such as polymers, nylon, or fiberglass, but it can also be made of polyethylene. It is usually a yellow or white paper-like sheet that has a glossy sheen. It comes in wide rolls with a width measuring up to 9 feet. As the name suggests, a house wrap functions more like gift-wrap protecting and sealing the item inside by covering all edges and functioning as a boxing cover around the house.

1. Virtually all houses have wood in their structure, moisture during the framing of the house is a dilemma you will face from time to time especially if you don't water-proof your house. When moisture seeps into it, it could rot, encourage mold growth, and cause health hazards to the occupants. House wrap can help prevent these things by protecting the structure and surface of your house from moisture damage. The plastic membrane of the house wrap acts as a barrier and prevents water seepage. As it keeps the structure dry inside, you will not have to worry about these problems.

2. Aside from shielding the house' structure from damaging buildup of moisture, a house wrap also serves as a good insulating material. The plastic-like, non-airtight feature of a house wrap limits air flow and reduces heat exchange from the interior and exterior of your home. This insulating effect, in turn, helps ease the burden on the part of your home conditioning system, helps you save energy, and lower your utility bills.

3. Not only does house wrap serve as a multi-purpose weather barrier, it also acts as a drainage plane. House wrap blocks water from passing through the structure but lets water vapor seep inside. This, then, allows moist humid air to break out from the inside while keeping liquid water off the interior of the home. Instead, it directs the rain water away from the house structure and straight to the ground.

Aside from these three main functions, another benefit of using house wrap is in terms of its durability. Although made of a thin sheet, house wrap can withstand normal wear and tear. If installed properly, it could last for a long time without showing signs of deterioration. However, according to Federal Emergency Management Administration or FEMA, compared to building paper, the latter can endure extreme hot or extreme cold climates and high winds better. Therefore, house wrap may not work well on places with these kinds of climate conditions.

Overall, house wrap is a great investment you should consider when constructing a new home. Though it has a number of advantages, installing it the wrong way can prove very hazardous rather than useful. FEMA has instructed that the sheets should be installed in horizontal rows with its top and bottom overlapping and seams covered with protective tape.

There are specialists who can do the job for you if you believe you are not up for the challenge. Proper installation is the key to making your house wrap work for you. If you are unsure how to go about installing house wrap, check with your local hardware store for some advice. Contractors in your area may also be able to come out to see the job and offer their expertise.

Basement Waterproofing

The basement waterproofing systems prevent water from leaking into the basement and damaging the foundation and wood. Basement waterproofing systems are needed whenever a basement, cellar, or other room is built at ground level or below. They are particularly important in areas where it is likely that ground water will raise the water table.

Without proper basement waterproofing systems, the basement can become cracked, and water pressure can result in serious damage to walls, as well as the growth mold and wood rot. Building the right kind of waterproofing system depends very much upon the environment in which the house is built.

Interior Sealant

Using interior sealants as basement waterproofing systems is a temporary measure, to be used during winter months to prevent snow and frost from raising the water table.

These sealants are often found in chemical spray form, and can be directly applied to walls and floors. Interior sealants can also prevent humidity and condensation within the basement. They can be absorbed by woods and porous building materials, causing cracks in masonry, damage to concrete, and wood rot. Masonry is also protected against spalling.

The most common use of these sealants is to prevent humidity inside the house from affecting the walls of the basement. For more secure measures against ground water, you will need to have either an exterior sealant or a drainage system.

Exterior Sealant

Exterior basement waterproofing systems stop ground water from reaching the basement walls as well as prevent mold and other damage which can occur in wet basement areas.

Waterproofing an exterior is the recognized IBC method to prevent damage caused by water. Exterior sealants were once only asphalt-based damp proofing, but now the most common kind of exterior basement waterproofing systems use a polymer base. This kind of material will last for the life of the building.

Other exterior sealants will probably erode in 20 years. You will not usually have to add anything to an exterior sealant, which should applied during the construction of the house, particularly in areas where flooding, heavy rain, or hurricanes are likely to occur.

Water Drainage

Drainage can be used to mitigate basement water and is often considered to be another form of basement waterproofing. Water drainage functions by drawing water away from the foundation, and forcing it into a drain, or through a pump system. Pump kids are available which can be installed in DIY form or by professional plumbers.

Water drainage basement waterproofing systems often need to be run on an isolated electric system in case of power shortage, especially during periods of storms. Sump pumps should be placed in a pit and sealed with a lid in order to keep the water away from the electricity. Doing so also prevents humidity in the pump from entering the atmosphere of the basement. Keeping the lid airtight ensures that poisonous gasses won't seep into the house.

Methods Used in Basement Waterproofing

Two methods used in basement waterproofing is the Positive(application on the same side of hydro-static pressure) and the Negative(application on the opposite side of water pressure) waterproofing.

- Positive Waterproofing: In this method, the surrounding perimeters to the building structure is dug – about a meter deep(depends) and layers of waterproofing solution is applied. This is the best suggested way to protect since the major cause of problem is water pressure from the outside during monsoons.
- Negative Waterproofing: This method must be used in extreme conditions when there is no other way of performing the task from the outside. Cautions must be taken that the materials used are not permeable and do not form honeycomb structures. This application is not very reliable because when the water seeps in, the proofing layer is not always flexible enough to adhere that much pressure. It has been a trend by many contractors to use this method because of the ease and faster application. But this must be avoided and Positive (outside) applications must be undertaken.
- Already Constructed basements can also apply a layer of Kota – Patthar/Stone tiles with Injection Grouting method.

A wet basement not only prevents you from enjoying additional space in your house but also can turn your basement into a giant petri dish perfect for growing unhealthy molds and fungi.

A Dry Basement is a Useable Basement

Installing a drain system is filthy, backbreaking work, but it's not complicated. With a little instruction from our drain tile experts, you can do a first-class job. And DIY pays off big: Pros charge $5,000 to $8,000 for a typical job (120 linear feet of drain tile). You can install yours for less than $1,500 in materials and tool rentals.

Rite-Way Waterproofing has been in the business of installing waterproofing basement drain systems since 1965. The pros there know what works, and they'd better, because the company has installed 100,000 systems, each with a lifetime guarantee. They also know what doesn't work—half of Rite-Way's current jobs involve replacing failed systems installed by other contractors.

A drain tile system creates a perfect pathway for dangerous radon gas to escape. If you've never tested for radon, it's smart to do so before you install a drain system. That way, you can plan for a radon mitigation system as well.

Before you Get Started

It's always best to stop water from entering your basement in the first place, so before you run to the rental center for your jackhammer, be sure to address the exterior issues. The grade next to the house should slope down away from the building by least 6 in. for the first 10 ft. Consider installing gutters, or make sure the existing gutters are working properly. And check that your irrigation system isn't adding to the problem by spraying water right up against the side of the house.

If your waterproofing basement is finished, with stud walls and insulation covering the foundation walls, you can still install a drain system. When you break out the concrete, leave small sections of floor intact so the wall doesn't drop down. A 4×4 in. section every 6 ft. is enough to support the wall. If there are obstacles along the wall (like a furnace), plan to tunnel under them. You'll find most of the materials you'll need at a home center. Order the rock from a landscape supplier. You'll also need a pickup to haul the dirt to the landfill.

Always check with your local building official. Explain your project, and see if any permits or inspections are required in your area. Sometimes, a building official who has been around for a while may have information on how your house was built or what issues you may run into in your area.

Control the Dust

Get Ready for Dust

Instead of just covering your stuff with sheets of plastic, isolate your work area with a wall of plastic sheeting. Make sure to fill in the spaces between joists.

There's no getting around it: Busting up concrete is a dirty job. Shut down your furnace or central air conditioning while you're working, and cover all return air vents until you're finished cleaning up. Instead of covering your furnishings with plastic, move everything out of the area and drape plastic from the ceiling to create an isolated work space. If you have an unfinished ceiling, be sure you run the plastic up into every joist space. Set a fan in the window to exhaust the heavy dust while you run the jackhammer. And wear a dust mask and hearing protection.

Bust up the Floor

Bust up the Floor.

Remove 16 in. to 18 in. of concrete along the wall with a rented electric jackhammer. Start by chipping in a straight line along the entire length of the wall, then come back and bust it into manageable chunks.

The pros use electric jackhammers, because the air that runs pneumatic jackhammers kicks up a lot more dust. You can get one from a rental center for under $100 a day. Start by hammering a line about 16 in. to 18 in. away from the wall.

Once the perimeter is done, come back and break the row of concrete into manageable chunks. Each section will break free easier if it has room to pull away, so remove the sections as you go. If you're working alone, make the most of your rental time; just set the chunks aside until you're done with the hammer. Don't forget to bust up a larger area for your sump basin.

Dig the Trench

Dig a trench along the wall.

Dig the trench as deep as the bottom of the footing. Instead of lugging pails of soil up the stairs, buy buckets that will fit through your basement windows.

Once the concrete is removed, dig down to the bottom of the footing but not below. If you compromise the soil under the footing, you could end up with cracks in your wall, or worse.

Five-gallon buckets are OK for hauling out debris, but the guys at Rite-Way use rubber feed buckets (typically used for farm animals) because they fit through small basement windows and are less likely to bang up trim. You can get them at farm supply retailers for under $20. And when it's time to haul the debris away, you may find that the landfill considers it to be "clean fill." You may be able to dump it for free.

Install the Basin

It's best to locate your basin in an unfinished area of the basement so you can have easy access to the sump pump. If you never plan on finishing the basement, locate the basin in the same area where you want the water to drain out of the house so you don't have as much plastic pipe to install. Dig the hole so the top of the basin will sit flush with the finished concrete.

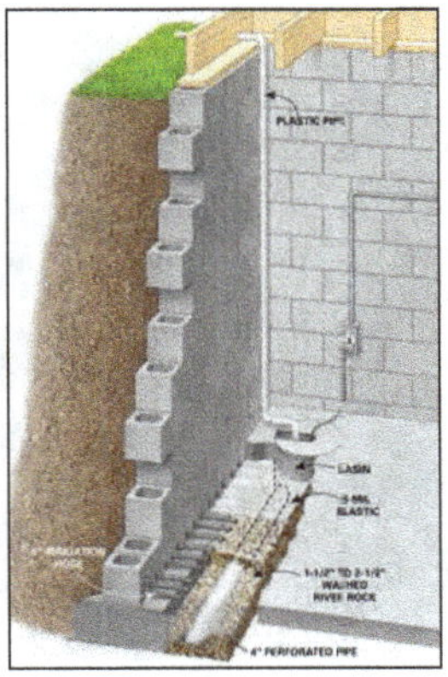

Many basins come with flat "knockout" areas meant to make cutting the hole easier. Don't assume the location of these knockouts will work for your system. Because the pipe will be slightly sloped down toward the basin, the longer the drain is, the lower the pipe will be when it reaches the basin. You never want standing water in your drainpipes, so make sure to choose a model that is deep enough. Our experts typically use 30-in.-deep basins. They use 36-in.-deep models for systems longer than 120 ft., and they install two basins if the drain is longer than 180 ft.

Set the basin in place, and then mark the locations for the holes where the pipes will meet the basin. Keep in mind there will be a thin layer of rock (one layer thick) under the pipe near the basin. Cut the holes using a reciprocating saw, jigsaw or hole saw. The holes don't have to be perfect.

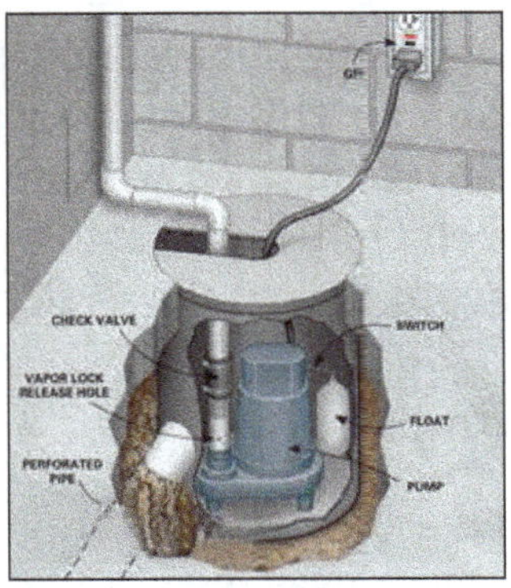

Don't haul out all the dirt right away; you'll need some to fill in around the basin. Once it's permanently in place, fill in around it, tamping the dirt with a 2×4 as you go.

Caution: Never drill holes in the bottom of the basin. If you have a high water table, water could come up from the bottom, and your pump will run nonstop, attempting to dry out the neighborhood.

Drill Holes in Block

If your waterproofing basement walls are made from concrete block, drill 1-in. holes in each block core and into each mortar joint. This will allow the water that collects in the cores and between the blocks to flow into the drain. Drill the holes as close to the footings as you can. You may find that the bottom blocks are filled with concrete. In that case, you'll have to remove any existing walls and install foundation wrap. Cut down on dust by laying a shop vacuum hose next to the hole as you drill.

Install the Pipe

Lay in the pipe.

Lay the pipe with the holes face down. Add or remove rock to slope the pipe so water flows to the sump basin. There's no need to cement the pipe connections

Before you lay the pipe in the trench, shovel in a bottom layer of 1-1/2-in. to 2-in. washed river rock (a layer of smaller rock can become clogged with minerals and sediments). The pipe should slope toward the basin at least 1/4 in. for every 10 ft. Rake the rock around to achieve this pitch.

Lay your irrigation pipe on top of the rock. Don't use ordinary flexible drainpipe because it clogs easily. The pros prefer to lay down a 4-in.-diameter Schedule 10 perforated pipe. Buy the kind of pipe with rows of 1/2-in. perforation holes only on one side, not all around the pipe.

Lay the pipe with the holes facing down, so the minerals and sediment in the water can flow down around the pipe and settle into the ground. This way, the water that does rise up into the pipes from underneath will be relatively clean. Clean water will add years to the life of the whole system. Start at the basin, and push the male end of the pipe into the basin about 4 in. Use PVC or ABS elbows at the corners. It's not necessary to cement the sections together.

Divert the Water into the Drain

Drill 1-in. holes in each block core and each mortar joint. Then insert sections of 1-in. irrigation hose from the holes into the gravel to carry away the water.

Once your pipe is installed, it's time to install the 1-in. irrigation hose that will carry the water from the blocks to the trench. Softer hose, like garden hose, can get crushed flat by the new concrete, so stick with irrigation hose. Cut the hose with a hacksaw or reciprocating saw. Make sure each section of hose runs several inches past the footing.

Install hose in the blcck walls.

If you have a Solid Concrete Foundation

If you have poured concrete walls rather than block, you'll need to install a foundation wrap to let the water into the drain as shown below. Foundation wrap is made from tough plastic and consists of rows of dimples that allow water to flow behind it. Platon by CertainTeed is one example. Home centers can order it if they don't stock it.

Cut the sheets into strips with a utility knife. Bend the strips at 90 degrees and let the bottom half run past the footing. The length of the wrap that's up against the wall depends on your situation. At a bare minimum, run the wrap up 4 in. above the top of the concrete on a poured wall, or up 4 in. past the holes you drilled in the block wall. If you're working along stud walls, try to tuck the wrap behind the bottom plate.

Cover it Back up

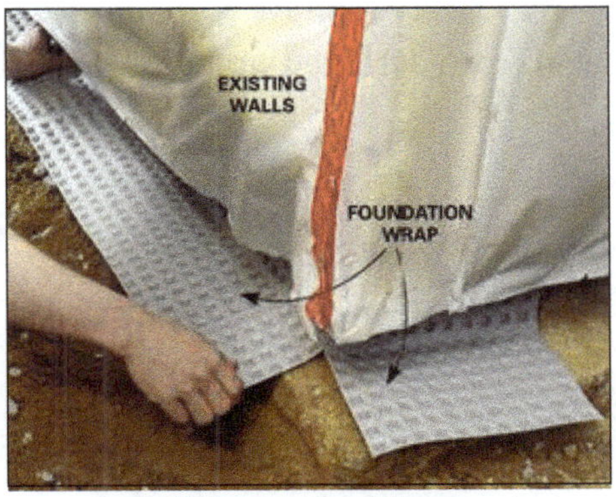

Cap off the trench with concrete.

Lay plastic over the rock and cover it with concrete. Smooth out the concrete with a float. Wait 20 minutes, and work it smooth with a steel trowel.

Once the hoses are in or the foundation wrap is in place, it's time to cover it up. Fill in the trench with river rock up to the bottom of the existing slab, and then cover the rock with at least a 6-mil thickness of plastic for a vapor barrier.

To cut down on dust, mix your concrete outside. Bagged concrete mix for slabs and sidewalks works just fine. Slide a 3-ft. section of 2×4 along the floor to "screed" the new concrete flush with the floor, and then smooth it out with a hand float. Wait 20 minutes, then smooth it with a finishing trowel. Use the float to completely fill the gap under any existing walls.

Hook up the Pump

Pros prefer submersible pumps that have a vertical float switch on them because they're more reliable than pedestal or float switch pumps. Install a 6- to 8-in. section of pipe on the pump, then a check valve. Make sure the check valve doesn't interfere with the pump switch. Above the check valve, attach another section of pipe long enough to reach above the top of the basin.

Drill a 1/4-in. to 3/8-in. vapor lock release hole in the section of pipe that's just below the check valve. This allows the pump to get up to speed before trying to force open the check valve, which may have many gallons of water pressing down on it. Angle the hole so water sprays down while the pump is working. It's best if you have a dedicated outlet for your pump. Extension cords get unplugged, and other appliances hooked up to the same circuit could trip a breaker.

The pipe that exits the basement needs to be located in an area that slopes away from the house. If that means running a pipe back across the basement, consider burying the waste pipe in the trench and having it come back up where you want it. If your pipe is going to discharge above grade and you live in a cold climate, run the pipe no more than 8 in. past the siding. This will keep it from freezing up in the winter.

With few exceptions, basement drain water cannot be dumped into city sewer systems. Most systems can be drained into storm sewers as long as they're above grade when they do. Ask your building official what the rules are in your area.

Required Tools for this Project

Have the necessary tools for this DIY project lined up before you start—you'll save time and frustration.

- Bucket
- Chalk line
- Cold chisel
- Dust mask
- Extension cord
- Hacksaw
- Hammer drill
- Jigsaw
- Knee pads
- Knockdown knife
- Reciprocating saw

- Safety glasses
- Shop vacuum
- Spade
- Tape measure
- Trowel
- Wheelbarrow

References

- Moisture-damage-465: comfortsolutions.ie, Retrieved 20 June 2020
- Tar-paper-facts-and-tips: askthebuilder.com, Retrieved 12 April 2020
- What-is-damp-proofing, damp-proofing: rentokil.co.uk, Retrieved 18 June 2020
- Building-envelope, encyclopedia: energyeducation.ca, Retrieved 25 April 2020
- Planning-24: fromplattoplace.com, Retrieved 26 March 2020
- Types-of-waterproofing-basement-waterproofing, home-improvement: broomberg.in, Retrieved 16 July 2020
- Breathability, how-to-deal-with-damp: Retrieved 22 March 2020

5

Sustainable Construction

The branch of civil engineering which aims to minimize the effect which civil engineering activities have upon the environment is termed as sustainable civil engineering. The following chapter elucidates the varied aspects associated with this area of study such as green building, groundwater banking, seawall and rain garden, etc.

Dam

A dam is a barrier that impounds water or underground streams. Reservoirs created by dams not only suppress floods but also provide water for such activities as irrigation, human consumption, industrial use, aquaculture, and navigability. Hydropower is often used in conjunction with dams to generate electricity. A dam can also be used to collect water or for storage of water which can be evenly distributed between locations. Dams generally serve the primary purpose of retaining water, while other structures such as floodgates or levees (also known as dikes) are used to manage or prevent water flow into specific land regions.

A sideview of the Lake Vyrnwy dam, in Wales, finished in 1888.

Karapuzha Dam, an earth dam in the Indian state of Kerala.

The word *dam* can be traced back to Middle English, and before that, from Middle Dutch, as seen in the names of many old cities.

History

Ancient Dams

Early dam building took place in Mesopotamia and the Middle East. Dams were used to control the water level, for Mesopotamia's weather affected the Tigris and Euphrates rivers.

The earliest known dam is the Jawa Dam in Jordan, 100 kilometres (62 mi) northeast of the capital Amman. This gravity dam featured an originally 9-metre-high (30 ft) and 1 m-wide (3.3 ft) stone wall, supported by a 50 m-wide (160 ft) earth rampart. The structure is dated to 3000 BC.

The Ancient Egyptian Sadd-el-Kafara Dam at Wadi Al-Garawi, located about 25 km (16 mi) south of Cairo, was 102 m (335 ft) long at its base and 87 m (285 ft) wide. The structure was built around 2800 or 2600 BC as a diversion dam for flood control, but was destroyed by heavy rain during construction or shortly afterwards. During the XIIth dynasty in the 19th century BC, the Pharaohs Senosert III, Amenemhat III and Amenmehat IV dug a canal 16 km (9.9 mi) long linking the Fayum Depression to the Nile in Middle Egypt. Two dams called Ha-Uar running east-west were built to retain water during the annual flood and then release it to surrounding lands. The lake called "Mer-wer" or Lake Moeris covered 1,700 km^2 (660 sq mi) and is known today as Berkat Qaroun.

One of the engineering wonders of the ancient world was the Great Dam of Marib in Yemen. Initiated somewhere between 1750 and 1700 BC, it was made of packed earth - triangular in cross section, 580 m (1,900 ft) in length and originally 4 m (13 ft) high

- running between two groups of rocks on either side, to which it was linked by substantial stonework. Repairs were carried out during various periods, most important around 750 BC, and 250 years later the dam height was increased to 7 m (23 ft). After the end of the Kingdom of Saba, the dam fell under the control of the Ḥimyarites (~115 BC) who undertook further improvements, creating a structure 14 m (46 ft) high, with five spillway channels, two masonry-reinforced sluices, a settling pond, and a 1,000 m (3,300 ft) canal to a distribution tank. These extensive works were not actually finalized until 325 AD and allowed the irrigation of 25,000 acres (100 km²).

By the mid-late 3rd century BC, an intricate water-management system within Dholavira in modern-day India was built. The system included 16 reservoirs, dams and various channels for collecting water and storing it.

Eflatun Pınar is a Hittite dam and spring temple near Konya, Turkey. It is thought to be from the time of the Hittite empire between the 15th and 13th century BC.

The Kallanai is constructed of unhewn stone, over 300 m (980 ft) long, 4.5 m (15 ft) high and 20 m (66 ft) wide, across the main stream of the Kaveri river in Tamil Nadu, South India. The basic structure dates to the 2nd century AD and is considered one of the oldest water-diversion or water-regulator structures in the world which is still in use. The purpose of the dam was to divert the waters of the Kaveri across the fertile delta region for irrigation via canals.

Du Jiang Yan is the oldest surviving irrigation system in China that included a dam that directed waterflow. It was finished in 251 BC. A large earthen dam, made by Sunshu Ao, the prime minister of Chu (state), flooded a valley in modern-day northern Anhui province that created an enormous irrigation reservoir (100 km (62 mi) in circumference), a reservoir that is still present today.

Roman Engineering

The Roman dam at Cornalvo in Spain has been in use for almost two millennia.

Roman dam construction was characterized by "the Romans' ability to plan and organize engineering construction on a grand scale." Roman planners introduced the then-novel concept of large reservoir dams which could secure a permanent water supply for urban settlements over the dry season. Their pioneering use of water-proof hydraulic mortar and particularly Roman concrete allowed for much larger dam structures than previously built, such as the Lake Homs Dam, possibly the largest water barrier to that date, and the Harbaqa Dam, both in Roman Syria. The highest Roman dam was the Subiaco Dam near Rome; its record height of 50 m (160 ft) remained unsurpassed until its accidental destruction in 1305.

Roman engineers made routine use of ancient standard designs like embankment dams and masonry gravity dams. Apart from that, they displayed a high degree of inventiveness, introducing most of the other basic dam designs which had been unknown until then. These include arch-gravity dams, arch dams, buttress dams and multiple arch buttress dams, all of which were known and employed by the 2nd century AD. Roman workforces also were the first to build dam bridges, such astheBridge of Valerian in Iran.

Remains of the Band-e Kaisar dam, built by the Romans in the 3rd century AD.

In Iran, bridge dams such as the Band-e Kaisar were used to provide hydropower through water wheels, which often powered water-raising mechanisms. One of the first was the Roman-built dam bridge in Dezful, which could raise water 50 cubits in height for the water supply to all houses in the town. Also diversion dams were known. Milling dams were introduced which the Muslim engineers called the *Pul-i-Bulaiti*. The first was built at Shustar on the River Karun, Iran, and many of these were later built in other parts of the Islamic world. Water was conducted from the back of the dam through a large pipe to drive a water wheel and watermill. In the 10th century, Al-Muqaddasi described several dams in Persia. He reported that one in Ahwaz was more than 910 m (3,000 ft) long, and that it had many water-wheels raising the water into aqueducts through which it flowed into reservoirs of the city. Another one, the Band-i-Amir dam, provided irrigation for 300 villages.

Middle Ages

In the Netherlands, a low-lying country, dams were often applied to block rivers in order to regulate the water level and to prevent the sea from entering the marsh lands. Such dams often marked the beginning of a town or city because it was easy to cross the river at such a place, and often gave rise to the respective place's names in Dutch.

For instance the Dutch capital Amsterdam (old name *Amstelredam*) started with a dam through the river Amstel in the late 12th century, and Rotterdam started with a dam through the river Rotte, a minor tributary of the Nieuwe Maas. The central square of Amsterdam, covering the original place of the 800-year-old dam, still carries the name *Dam Square* or simply *the Dam.*

Industrial Revolution

An engraving of the Rideau Canal locks at Bytown.

The Romans were the first to build arch dams, where the reaction forces from the abutment stabilizes the structure from the external hydrostatic pressure, but it was only in the 19th century that the engineering skills and construction materials available were capable of building the first large-scale arch dams.

Three pioneering arch dams were built around the British Empire in the early 19th century. Henry Russel of the Royal Engineers oversaw the construction of the Mir Alam dam in 1804 to supply water to the city of Hyderabad (it is still in use today). It had a height of 12 m (39 ft) and consisted of 21 arches of variable span.

In the 1820s and 30s, Lieutenant-Colonel John By supervised the construction of the Rideau Canal in Canada near modern-day Ottawa and built a series of curved masonry dams as part of the waterway system. In particular, the Jones Falls Dam, built by John Redpath, was completed in 1832 as the largest dam in North America and an engineering marvel. In order to keep the water in control during construction, two sluices, artificial channels for conducting water, were kept open in the dam. The first was near the base of the dam on its east side. A second sluice was put in on the west side of the dam, about 20 ft (6.1 m) above the base. To make the switch from the lower to upper sluice, the outlet of Sand Lake was blocked off.

Masonry arch wall, Parramatta, New South Wales, the first engineered dam built in Australia.

Hunts Creek near the city of Parramatta, Australia, was dammed in the 1850s, to cater for the demand for water from the growing population of the city. The masonry arch dam wall was designed by Lieutenant Percy Simpson who was influenced by the advances in dam engineering techniques made by the Royal Engineers in India. The dam cost £17,000 and was completed in 1856 as the first engineered dam built in Australia, and the second arch dam in the world built to mathematical specifications.

The first such dam was opened two years earlier in France. It was the first French arch dam of the industrial era, and it was built by François Zola in the municipality of Aix-en-Provence to improve the supply of water after the 1832 cholera outbreak devastated the area. After royal approval was granted in 1844, the dam was constructed over the following decade. Its construction was carried out on the basis of the mathematical results of scientific stress analysis.

The 75-miles dam near Warwick, Australia, was possibly the world's first concrete arch dam. Designed by Henry Charles Stanley in 1880 with an overflow spillway and a special water outlet, it was eventually heightened to 10 m (33 ft).

In the latter half of the nineteenth century, significant advances in the scientific theory of masonry dam design were made. This transformed dam design from an art based on empirical methodology to a profession based on a rigorously applied scientific theoretical framework. This new emphasis was centered around the engineering faculties of universities in France and in the United Kingdom. William John Macquorn Rankine at the University of Glasgow pioneered the theoretical understanding of dam structures in his 1857 paper *On the Stability of Loose Earth*. Rankine theory provided a good understanding of the principles behind dam design. In France, J. Augustin Tortene de Sazilly explained the mechanics of vertically faced masonry gravity dams, and Zola's dam was the first to be built on the basis of these principles.

Large Dams

The Hoover Dam by Ansel Adams, 1942.

The era of large dams was initiated with the construction of the Aswan Low Dam in Egypt in 1902, a gravity masonry buttress dam on the Nile River. Following their 1882 invasion and occupation of Egypt, the British began construction in 1898. The project was designed by Sir William Willcocks and involved several eminent engineers of the time, including Sir Benjamin Baker and Sir John Aird, whose firm, John Aird & Co., was the main contractor. Capital and financing were furnished by Ernest Cassel. When initially constructed between 1899 and 1902, nothing of its scale had ever been attempted; on completion, it was the largest masonry dam in the world.

The Hoover Dam is a massive concrete arch-gravity dam, constructed in the Black Canyon of the Colorado River, on the border between the US states of Arizona and Nevada between 1931 and 1936 during the Great Depression. In 1928, Congress authorized the project to build a dam that would control floods, provide irrigation water and produce hydroelectric power. The winning bid to build the dam was submitted by a consortium called Six Companies, Inc. Such a large concrete structure had never been built before, and some of the techniques were unproven. The torrid summer weather and the lack of facilities near the site also presented difficulties. Nevertheless, Six Companies turned over the dam to the federal government on 1 March 1936, more than two years ahead of schedule.

By 1997, there were an estimated 800,000 dams worldwide, some 40,000 of them over 15 m (49 ft) high. In 2014, scholars from the University of Oxford published a study of the cost of large dams – based on the largest existing dataset – documenting significant cost overruns for a majority of dams and questioning whether benefits typically offset costs for such dams.

Types of Dams

Dams can be formed by human agency, natural causes, or even by the intervention of wildlife such as beavers. Man-made dams are typically classified according to their size (height), intended purpose or structure.

By Structure

Based on structure and material used, dams are classified as easily created without materials, arch-gravity dams, embankment dams or masonry dams, with several subtypes.

Arch Dams

Gordon Dam, Tasmania, is an arch dam.

In the arch dam, stability is obtained by a combination of arch and gravity action. If the upstream face is vertical the entire weight of the dam must be carried to the foundation by gravity, while the distribution of the normal hydrostatic pressure between vertical cantilever and arch action will depend upon the stiffness of the dam in a vertical and horizontal direction. When the upstream face is sloped the distribution is more complicated. The normal component of the weight of the arch ring may be taken by the arch action, while the normal hydrostatic pressure will be distributed as described above.

For this type of dam, firm reliable supports at the abutments (either buttress or canyon side wall) are more important. The most desirable place for an arch dam is a narrow canyon with steep side walls composed of sound rock. The safety of an arch dam is dependent on the strength of the side wall abutments, hence not only should the arch be well seated on the side walls but also the character of the rock should be carefully inspected.

Daniel-Johnson Dam, Quebec, is a multiple-arch buttress dam.

Two types of single-arch dams are in use, namely the constant-angle and the constant-radius dam. The constant-radius type employs the same face radius at all elevations of the dam, which means that as the channel grows narrower towards the bottom of the dam the central angle subtended by the face of the dam becomes smaller. Jones Falls Dam, in Canada, is a constant radius dam. In a constant-angle dam, also known as a variable radius dam, this subtended angle is kept a constant and the variation in distance between the abutments at various levels are taken care of by varying the radii. Constant-radius dams are much less common than constant-angle dams. Parker Dam on the Colorado River is a constant-angle arch dam.

A similar type is the double-curvature or thin-shell dam. Wildhorse Dam near Mountain City, Nevada, in the United States is an example of the type. This method of construction minimizes the amount of concrete necessary for construction but transmits large loads to the foundation and abutments. The appearance is similar to a single-arch dam but with a distinct vertical curvature to it as well lending it the vague appearance of a concave lens as viewed from downstream.

The multiple-arch dam consists of a number of single-arch dams with concrete buttresses as the supporting abutments, as for example the Daniel-Johnson Dam, Québec, Canada. The multiple-arch dam does not require as many buttresses as the hollow gravity type, but requires good rock foundation because the buttress loads are heavy.

Gravity Dams

The Grand Coulee Dam is an example of a solid gravity dam.

In a gravity dam, the force that holds the dam in place against the push from the water is Earth's gravity pulling down on the mass of the dam. The water presses laterally (downstream) on the dam, tending to overturn the dam by rotating about its toe (a point at the bottom downstream side of the dam). The dam's weight counteracts that force, tending to rotate the dam the other way about its toe. The designer ensures that the dam is heavy enough that the dam's weight wins that contest. In engineering terms, that is true whenever the resultant of the forces of gravity acting on the dam and water pressure on the dam acts in a line that passes upstream of the toe of the dam.

Furthermore, the designer tries to shape the dam so if one were to consider the part of dam above any particular height to be a whole dam itself, that dam also would be held in place by gravity. i.e. there is no tension in the upstream face of the dam holding the top of the dam down. The designer does this because it is usually more practical to make a dam of material essentially just piled up than to make the material stick together against vertical tension.

Note that the shape that prevents tension in the upstream face also eliminates a balancing compression stress in the downstream face, providing additional economy.

For this type of dam, it is essential to have an impervious foundation with high bearing strength.

When situated on a suitable site, a gravity dam can prove to be a better alternative to other types of dams. When built on a carefully studied foundation, the gravity dam probably represents the best developed example of dam building. Since the fear of flood is a strong motivator in many regions, gravity dams are being built in some instances where an arch dam would have been more economical.

Gravity dams are classified as "solid" or "hollow" and are generally made of either concrete or masonry. The solid form is the more widely used of the two, though the hollow dam is frequently more economical to construct. Grand Coulee Dam is a solid gravity dam and Braddock Locks & Dam is a hollow gravity dam.

Arch-Gravity Dams

The Hoover Dam is an example of an arch-gravity dam.

A gravity dam can be combined with an arch dam into an arch-gravity dam for areas with massive amounts of water flow but less material available for a purely gravity dam. The inward compression of the dam by the water reduces the lateral (horizontal) force acting on the dam. Thus, the gravitation force required by the dam is lessened, i.e. the dam does not need to be so massive. This enables thinner dams and saves resources.

Barrages

The Koshi Barrage.

A barrage dam is a special kind of dam which consists of a line of large gates that can be opened or closed to control the amount of water passing the dam. The gates are set between flanking piers which are responsible for supporting the water load, and are often used to control and stabilize water flow for irrigation systems. An example of this type of dam is the now-decommissioned Red Bluff Diversion Dam on the Sacramento River near Red Bluff, California.

Barrages that are built at the mouths of rivers or lagoons to prevent tidal incursions or utilize the tidal flow for tidal power are known as tidal barrages.

Embankment Dams

Embankment dams are made from compacted earth, and have two main types, rock-fill and earth-fill dams. Embankment dams rely on their weight to hold back the force of water, like gravity dams made from concrete.

Rock-Fill Dams

The Gathright Dam in Virginia is a rock-fill embankment dam.

Rock-fill dams are embankments of compacted free-draining granular earth with an impervious zone. The earth utilized often contains a high percentage of large particles, hence the term "rock-fill". The impervious zone may be on the upstream face and made of masonry, concrete, plastic membrane, steel sheet piles, timber or other material. The impervious zone may also be within the embankment in which case it is referred to as a *core*. In the instances where clay is utilized as the impervious material the dam is referred to as a *composite* dam. To prevent internal erosion of clay into the rock fill due to seepage forces, the core is separated using a filter. Filters are specifically graded soil designed to prevent the migration of fine grain soil particles. When suitable material is at hand, transportation is minimized leading to cost savings during construction. Rock-fill dams are resistant to damage from earthquakes. However, inadequate quality control during construction can lead to poor compaction and sand in the embankment which can lead to liquefaction of the rock-fill during an earthquake. Liquefaction potential can be reduced by keeping susceptible material from being saturated, and by providing adequate compaction during construction. An example of a rock-fill dam is New Melones Dam in California or the Fierza Dam in Albania.

A core that is growing in popularity is asphalt concrete. The majority of such dams are built with rock and/or gravel as the main fill material. Almost 100 dams of this design have now been built worldwide since the first such dam was completed in 1962. All asphalt-concrete core dams built so far have an excellent performance record. The type of asphalt used is a viscoelastic-plastic material that can adjust to the movements and deformations imposed on the embankment as a whole, and to settlements in the foundation. The flexible properties of the asphalt make such dams especially suited in earthquake regions.

For the Moglicë Hydro Power Plant in Albania the Norwegian power company Statkraft is currently building an asphalt-core rock-fill dam. Upon completion in 2018 the 320 m long, 150 m high and 460 m wide dam is anticipated to be the world's highest of its kind.

Concrete-Face Rock-Fill Dams

A concrete-face rock-fill dam (CFRD) is a rock-fill dam with concrete slabs on its upstream face. This design provides the concrete slab as an impervious wall to prevent leakage and also a structure without concern for uplift pressure. In addition, the CFRD design is flexible for topography, faster to construct and less costly than earth-fill dams. The CFRD concept originated during the California Gold Rush in the 1860s when miners constructed rock-fill timber-face dams for sluice operations. The timber was later replaced by concrete as the design was applied to irrigation and power schemes. As CFRD designs grew in height during the 1960s, the fill was compacted and the slab's horizontal and vertical joints were replaced with improved vertical joints. In the last few decades, the design has become popular.

Currently, the tallest CFRD in the world is the 233 m-tall (764 ft) Shuibuya Dam in China which was completed in 2008.

Earth-Fill Dams

Earth-fill dams, also called earthen dams, rolled-earth dams or simply earth dams, are constructed as a simple embankment of well compacted earth. A homogeneous rolled-earth dam is entirely constructed of one type of material but may contain a drain layer to collect seep water. A zoned-earth dam has distinct parts or zones of dissimilar material, typically a locally plentiful shell with a watertight clay core. Modern zoned-earth embankments employ filter and drain zones to collect and remove seep water and preserve the integrity of the downstream shell zone. An outdated method of zoned earth dam construction utilized a hydraulic fill to produce a watertight core. Rolled-earth dams may also employ a watertight facing or core in the manner of a rock-fill dam. An interesting type of temporary earth dam occasionally used in high latitudes is the frozen-core dam, in which a coolant is circulated through pipes inside the dam to maintain a watertight region of permafrost within it.

Tarbela Dam is a large dam on the Indus River in Pakistan. It is located about 50 km (31 mi) northwest of Islamabad, and a height of 485 ft (148 m) above the river bed and a reservoir size of 95 sq mi (250 km^2) makes it the largest earth-filled dam in the world. The principal element of the project is an embankment 9,000 feet (2,700 m) long with a maximum height of 465 feet (142 m). The total volume of earth and rock used for the project is approximately 200 million cubic yards (152.8 million cu. meters) which makes it one of the largest man-made structures in the world.

Because earthen dams can be constructed from materials found on-site or nearby, they can be very cost-effective in regions where the cost of producing or bringing in concrete would be prohibitive.

By Size

International standards (including the International Commission on Large Dams, ICOLD) define *large dams* as higher than 15 m (49 ft) and *major dams* as over 150 m (490 ft) in height. The *Report of the World Commission on Dams* also includes in the *large* category, dams, such as barrages, which are between 5 and 15 m (16 and 49 ft) high with a reservoir capacity of more than 3 million cubic metres (2,400 acre·ft).

The tallest dam in the world is the 300 m-high (980 ft) Nurek Dam in Tajikistan.

By Use

Saddle Dam

A saddle dam is an auxiliary dam constructed to confine the reservoir created by a primary dam either to permit a higher water elevation and storage or to limit the extent of a reservoir for increased efficiency. An auxiliary dam is constructed in a low spot or "saddle" through which the reservoir would otherwise escape. On occasion, a reservoir is contained by a similar structure called a dike to prevent inundation of nearby land. Dikes are commonly used for reclamation of arable land from a shallow lake. This is similar to a levee, which is a wall or embankment built along a river or stream to protect adjacent land from flooding.

Weir

A weir (also sometimes called an *overflow dam*) is a type of small overflow dam that is often used within a river channel to create an impoundment lake for water abstraction purposes and which can also be used for flow measurement or retardation.

Check Dam

A check dam is a small dam designed to reduce flow velocity and control soil erosion. Conversely, a wing dam is a structure that only partly restricts a waterway, creating a faster channel that resists the accumulation of sediment.

Dry Dam

A dry dam, also known as a flood retarding structure, is a dam designed to control flooding. It normally holds back no water and allows the channel to flow freely, except during periods of intense flow that would otherwise cause flooding downstream.

Diversionary Dam

A diversionary dam is a structure designed to divert all or a portion of the flow of a river from its natural course. The water may be redirected into a canal or tunnel for irrigation and/or hydroelectric power production.

Underground Dam

Underground dams are used to trap groundwater and store all or most of it below the surface for extended use in a localized area. In some cases they are also built to prevent saltwater from intruding into a freshwater aquifer. Underground dams are typically constructed in areas where water resources are minimal and need to be efficiently stored, such as in deserts and on islands like the Fukuzato Dam in Okinawa, Japan. They are most common in northeastern Africa and the arid areas of Brazil while also being used in the southwestern United States, Mexico, India, Germany, Italy, Greece, France and Japan.

There are two types of underground dams: a *sub-surface* and a *sand-storage* dam. A sub-surface dam is built across an aquifer or drainage route from an impervious layer (such as solid bedrock) up to just below the surface. They can be constructed of a variety of materials to include bricks, stones, concrete, steel or PVC. Once built, the water stored behind the dam raises the water table and is then extracted with wells. A sand-storage dam is a weir built in stages across a stream or wadi. It must be strong, as floods will wash over its crest. Over time, sand accumulates in layers behind the dam, which helps store water and, most importantly, prevent evaporation. The stored water can be extracted with a well, through the dam body, or by means of a drain pipe.

Tailings Dam

A tailings dam is typically an earth-fill embankment dam used to store tailings, which are produced during mining operations after separating the valuable fraction from the uneconomic fraction of an ore. Conventional water retention dams can serve this purpose, but due to cost, a tailings dam is more viable. Unlike water retention dams, a tailings dam is raised in succession throughout the life of the particular mine. Typically, a base or starter dam is constructed, and as it fills with a mixture of tailings and water, it is raised. Material used to raise the dam can include the tailings (depending on their size) along with dirt.

There are three raised tailings dam designs, the *upstream*, *downstream* and *centerline*, named according to the movement of the crest during raising. The specific design used is dependent upon topography, geology, climate, the type of tailings, and cost. An upstream tailings dam consists of trapezoidal embankments being constructed on top but toe to crest of another, moving the crest further upstream. This creates a relatively flat downstream side and a jagged upstream side which is supported by tailings slurry in the impoundment. The downstream design refers to the successive raising of the embankment that positions the fill and crest further downstream. A centerlined dam has sequential embankment dams constructed directly on top of another while fill is placed on the downstream side for support and slurry supports the upstream side.

Because tailings dams often store toxic chemicals from the mining process, they have an impervious liner to prevent seepage. Water/slurry levels in the tailings pond must be managed for stability and environmental purposes as well.

By Material

Steel Dams

Red Ridge steel dam, built 1905, Michigan.

A steel dam is a type of dam briefly experimented with around the start of the 20th century which uses steel plating (at an angle) and load-bearing beams as the structure. Intended as permanent structures, steel dams were an (arguably failed) experiment to determine if a construction technique could be devised that was cheaper than masonry, concrete or earthworks, but sturdier than timber crib dams.

Timber Dams

A timber crib dam in Michigan, photographed in 1978.

Timber dams were widely used in the early part of the industrial revolution and in frontier areas due to ease and speed of construction. Rarely built in modern times because of their relatively short lifespan and the limited height to which they can be built, timber dams must be kept constantly wet in order to maintain their water retention properties and limit deterioration by rot, similar to a barrel. The locations where timber dams are most economical to build are those where timber is plentiful, cement is costly or difficult to transport, and either a low head diversion dam is required or longevity is not an issue. Timber dams were once numerous, especially in the North American West, but most have failed, been hidden under earth embankments, or been replaced with entirely new structures. Two common variations of timber dams were the *crib* and the *plank*.

Timber crib dams were erected of heavy timbers or dressed logs in the manner of a log house and the interior filled with earth or rubble. The heavy crib structure supported the dam's face and the weight of the water. Splash dams were timber crib dams used to help float logs downstream in the late 19th and early 20th centuries.

Timber plank dams were more elegant structures that employed a variety of construction methods utilizing heavy timbers to support a water retaining arrangement of planks.

Other Types

Cofferdams

A cofferdam during the construction of locks at the Montgomery Point Lock and Dam.

A cofferdam is a barrier, usually temporary, constructed to exclude water from an area that is normally submerged. Made commonly of wood, concrete, or steel sheet piling, cofferdams are used to allow construction on the foundation of permanent dams, bridges, and similar structures. When the project is completed, the cofferdam will usually be demolished or removed unless the area requires continuous maintenance.

Common uses for cofferdams include construction and repair of offshore oil platforms. In such cases the cofferdam is fabricated from sheet steel and welded into place under water. Air is pumped into the space, displacing the water and allowing a dry work environment below the surface.

Natural Dams

Dams can also be created by natural geological forces. Volcanic dams are formed when lava flows, often basaltic, intercept the path of a stream or lake outlet, resulting in the creation of a natural impoundment. An example would be the eruptions of the Uinkaret volcanic field about 1.8 million–10,000 years ago, which created lava dams on the Colorado River in northern Arizona in the United States. The largest such lake grew to about 800 km (500 mi) in length before the failure of its dam. Glacial activity can also form natural dams, such as the damming of the Clark Fork in Montana by the Cordilleran Ice Sheet, which formed the 7,780 km^2 (3,000 sq mi) Glacial Lake Missoula near the end of the last Ice Age. Moraine deposits left behind by glaciers can also dam rivers to form lakes, such as at Flathead Lake, also in Montana.

Natural disasters such as earthquakes and landslides frequently create landslide dams in mountainous regions with unstable local geology. Historical examples include the Usoi Dam in Tajikistan, which blocks the Murghab River to create Sarez Lake. At 560 m (1,840 ft) high, it is the tallest dam in the world, including both natural and man-made dams. A more recent example would be the creation of Attabad Lake by a landslide on Pakistan's Hunza River.

Natural dams often pose significant hazards to human settlements and infrastructure. The resulting lakes often flood inhabited areas, while a catastrophic failure of the dam could cause even greater damage, such as the failure of western Wyoming's Gros Ventre landslide dam in 1927, which wiped out the town of Kelly and resulted in the deaths of six people.

Beaver Dams

Beavers create dams primarily out of mud and sticks to flood a particular habitable area. By flooding a parcel of land, beavers can navigate below or near the surface and remain relatively well hidden or protected from predators. The flooded region also allows beavers access to food, especially during the winter.

Construction Elements

Power Generation Plant

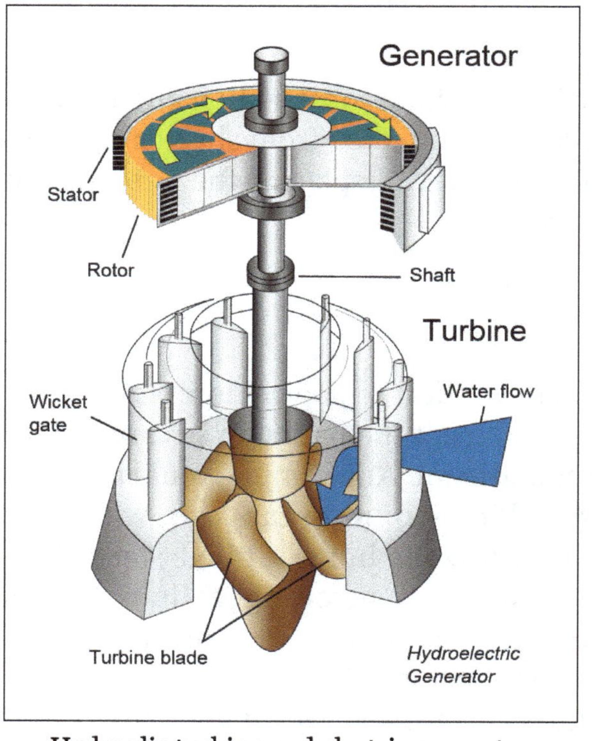

Hydraulic turbine and electric generator.

As of 2005, hydroelectric power, mostly from dams, supplies some 19% of the world's electricity, and over 63% of renewable energy. Much of this is generated by large dams, although China uses small-scale hydro generation on a wide scale and is responsible for about 50% of world use of this type of power.

Most hydroelectric power comes from the potential energy of dammed water driving a water turbine and generator; to boost the power generation capabilities of a dam, the water may be run through a large pipe called a penstock before the turbine. A variant on this simple model uses pumped-storage hydroelectricity to produce electricity to match periods of high and low demand, by moving water between reservoirs at different elevations. At times of low electrical demand, excess generation capacity is used to pump water into the higher reservoir. When there is higher demand, water is released back into the lower reservoir through a turbine.

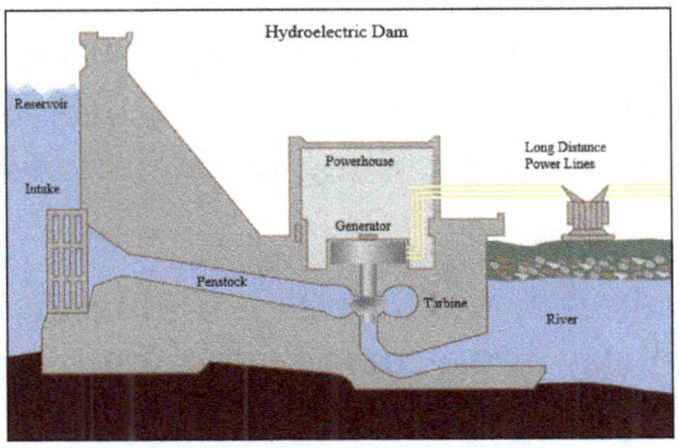

Hydroelectric dam in cross section.

Spillways

Spillway on Llyn Brianne dam, Wales, soon after first fill.

A spillway is a section of a dam designed to pass water from the upstream side of a dam to the downstream side. Many spillways have floodgates designed to control the flow through the spillway. There are several types of spillway. A *service spillway* or *primary spillway* passes normal flow. An *auxiliary spillway* releases flow in excess of the capacity of the service spillway. An *emergency spillway* is designed for extreme conditions, such as a serious malfunction of the service spillway. A *fuse plug spillway* is a low embankment designed to be overtopped and washed away in the event of a large flood. The elements of a fuse plug are independent free-standing blocks, set side by side which work without any remote control. They allow increasing the normal pool of the dam without compromising the security of the dam because they are designed to be gradually evacuated for exceptional events. They work as fixed weirs at times by allowing over-flow for common floods.

The spillway can be gradually eroded by water flow, including cavitation or turbulence of the water flowing over the spillway, leading to its failure. It was the inadequate design of the spillway which led to the 1889 over-topping of the South Fork Dam in Johnstown, Pennsylvania, resulting in the infamous Johnstown Flood (the "great flood of 1889").

Erosion rates are often monitored, and the risk is ordinarily minimized, by shaping the downstream face of the spillway into a curve that minimizes turbulent flow, such as an ogee curve.

Dam Creation

Common Purposes

Function	Example
Power generation	Hydroelectric power is a major source of electricity in the world. Many countries have rivers with adequate water flow, that can be dammed for power generation purposes. For example, the Itaipu Dam on the Paraná River in South America generates 14 GW and supplied 93% of the energy consumed by Paraguay and 20% of that consumed by Brazil as of 2005.
Water supply	Many urban areas of the world are supplied with water abstracted from rivers pent up behind low dams or weirs. Examples include London, with water from the River Thames, and Chester, with water taken from the River Dee. Other major sources include deep upland reservoirs contained by high dams across deep valleys, such as the Claerwen series of dams and reservoirs.
Stabilize water flow / irrigation	Dams are often used to control and stabilize water flow, often for agricultural purposes and irrigation. Others such as the Berg Strait dam can help to stabilize or restore the water levels of inland lakes and seas, in this case the Aral Sea.
Flood prevention	The Keenleyside Dam on the Columbia River, Canada can store 8.76 km^3 (2.10 cu mi) of floodwaters, and the huge Delta Works protects the Netherlands from coastal flooding.

Land reclamation	Dams (often called dykes or levees in this context) are used to prevent ingress of water to an area that would otherwise be submerged, allowing its reclamation for human use.
Water diversion	A typically small dam used to divert water for irrigation, power generation, or other uses, with usually no other function. Occasionally, they are used to divert water to another drainage or reservoir to increase flow there and improve water use in that particular area.
Navigation	Dams create deep reservoirs and can also vary the flow of water downstream. This can in return affect upstream and downstream navigation by altering the river's depth. Deeper water increases or creates freedom of movement for water vessels. Large dams can serve this purpose, but most often weirs and locks are used.

Some of these purposes are conflicting, and the dam operator needs to make dynamic tradeoffs. For example, power generation and water supply would keep the reservoir high, whereas flood prevention would keep it low. Many dams in areas where precipitation fluctuates in an annual cycle will also see the reservoir fluctuate annually in an attempt to balance these difference purposes. Dam management becomes a complex exercise amongst competing stakeholders.

Location

The discharge of Takato Dam.

One of the best places for building a dam is a narrow part of a deep river valley; the valley sides then can act as natural walls. The primary function of the dam's structure is to fill the gap in the natural reservoir line left by the stream channel. The sites are usually those where the gap becomes a minimum for the required storage capacity. The most economical arrangement is often a composite structure such as a masonry dam flanked by earth embankments. The current use of the land to be flooded should be dispensable.

Significant other engineering and engineering geology considerations when building a dam include:

- Permeability of the surrounding rock or soil.
- Earthquake faults.
- Landslides and slope stability.
- Water table.
- Peak flood flows.
- Reservoir silting.
- Environmental impacts on river fisheries, forests and wildlife.
- Impacts on human habitations.
- Compensation for land being flooded as well as population resettlement.
- Removal of toxic materials and buildings from the proposed reservoir area.

Impact Assessment

Impact is assessed in several ways: the benefits to human society arising from the dam (agriculture, water, damage prevention and power), harm or benefit to nature and wildlife, impact on the geology of an area (whether the change to water flow and levels will increase or decrease stability), and the disruption to human lives (relocation, loss of archeological or cultural matters underwater).

Environmental Impact

Wood and garbage accumulation due to a dam.

Reservoirs held behind dams affect many ecological aspects of a river. Rivers topography and dynamics depend on a wide range of flows, whilst rivers below dams often experience long periods of very stable flow conditions or sawtooth flow patterns caused by releases followed by no releases. Water releases from a reservoir including that exiting a turbine usually contain very little suspended sediment, and this in turn can lead to scouring of river beds and loss of riverbanks; for example, the daily cyclic flow variation caused by the Glen Canyon Dam was a contributor to sand bar erosion.

Older dams often lack a fish ladder, which keeps many fish from moving upstream to their natural breeding grounds, causing failure of breeding cycles or blocking of migration paths. Even the presence of a fish ladder does not always prevent a reduction in fish reaching the spawning grounds upstream. In some areas, young fish ("smolt") are transported downstream by barge during parts of the year. Turbine and power-plant designs that have a lower impact upon aquatic life are an active area of research.

A large dam can cause the loss of entire ecospheres, including endangered and undiscovered species in the area, and the replacement of the original environment by a new inland lake.

Large reservoirs formed behind dams have been indicated in the contribution of seismic activity, due to changes in water load and/or the height of the water table.

Dams are also found to have a role in the increase/decrease of global warming. The changing water levels in reservoirs are a source for greenhouse gases like methane. While dams and the water behind them cover only a small portion of earth's surface, they harbour biological activity that can produce large amounts of greenhouse gases.

Human Social Impact

The impact on human society is also significant. Nick Cullather argues in *Hungry World: America's Cold War Battle Against Poverty in Asia* that dam construction requires the state to displace individual people in the name of the common good, and that it often leads to abuses of the masses by planners. He cites Morarji Desai, Interior Minister of India, in 1960 speaking to villagers upset about the Pong Dam, who threatened to "release the waters" and drown the villagers if they did not cooperate.

For example, the Three Gorges Dam on the Yangtze River in China is more than five times the size of the Hoover Dam (U.S.), and creates a reservoir 600 km (370 mi) long to be used for flood control and hydro-power generation. Its construction required the loss of over a million people's homes and their mass relocation, the loss of many valuable archaeological and cultural sites, as well as significant ecological change. During the 2010 China floods, the dam held back a what would have been a disastrous flood and the huge reservoir rose by 4 m (13 ft) overnight.

It is estimated that to date, 40–80 million people worldwide have been physically displaced from their homes as a result of dam construction.

Economics

Construction of a hydroelectric plant requires a long lead time for site studies, hydrological studies, and environmental impact assessments, and are large-scale projects by comparison to traditional power generation based upon fossil fuels. The number of sites that can be economically developed for hydroelectric production is limited; new sites tend to be far from population centers and usually require extensive power transmission lines. Hydroelectric generation can be vulnerable to major changes in the climate, including variations in rainfall, ground and surface water levels, and glacial melt, causing additional expenditure for the extra capacity to ensure sufficient power is available in low-water years.

Once completed, if it is well designed and maintained, a hydroelectric power source is usually comparatively cheap and reliable. It has no fuel and low escape risk, and as an alternative energy source it is cheaper than both nuclear and wind power. It is more easily regulated to store water as needed and generate high power levels on demand compared to wind power.

Dam Failure

The reservoir emptying through the failed Teton Dam.

International special sign for works and installations containing dangerous forces.

Dam failures are generally catastrophic if the structure is breached or significantly damaged. Routine deformation monitoring and monitoring of seepage from drains in and around larger dams is useful to anticipate any problems and permit remedial action to be taken before structural failure occurs. Most dams incorporate mechanisms to permit the reservoir to be lowered or even drained in the event of such problems. Another solution can be rock grouting – pressure pumping portland cement slurry into weak fractured rock.

During an armed conflict, a dam is to be considered as an "installation containing dangerous forces" due to the massive impact of a possible destruction on the civilian population and the environment. As such, it is protected by the rules of international humanitarian law (IHL) and shall not be made the object of attack if that may cause severe losses among the civilian population. To facilitate the identification, a protective sign consisting of three bright orange circles placed on the same axis is defined by the rules of IHL.

The main causes of dam failure include inadequate spillway capacity, piping through the embankment, foundation or abutments, spillway design error (South Fork Dam), geological instability caused by changes to water levels during filling or poor surveying (Vajont, Malpasset, Testalinden Creek dams), poor maintenance, especially of outlet pipes (Lawn Lake Dam, Val di Stava Dam collapse), extreme rainfall (Shakidor Dam), earthquakes, and human, computer or design error (Buffalo Creek Flood, Dale Dike Reservoir, Taum Sauk pumped storage plant).

A notable case of deliberate dam failure (prior to the above ruling) was the Royal Air Force 'Dambusters' raid on Germany in World War II (codenamed "Operation Chastise"), in which three German dams were selected to be breached in order to damage German infrastructure and manufacturing and power capabilities deriving from the Ruhr and Eder rivers. This raid later became the basis for several films.

Since 2007, the Dutch IJkdijk foundation is developing, with an open innovation model and early warning system for levee/dike failures. As a part of the development effort, full-scale dikes are destroyed in the IJkdijk fieldlab. The destruction process is monitored by sensor networks from an international group of companies and scientific institutions.

Green Building

US EPA Kansas City Science & Technology Center This facility features the following green attributes:

- LEED 2.0 Gold certified
- Green Power
- Native Landscaping

Green building (also known as green construction or sustainable building) refers to both a structure and the using of processes that are environmentally responsible and resource-efficient throughout a building's life-cycle: from siting to design, construction, operation, maintenance, renovation, and demolition. In other words, green building design involves finding the balance between homebuilding and the sustainable environment. This requires close cooperation of the design team, the architects, the engineers, and the client at all project stages. The Green Building practice expands and complements the classical building design concerns of economy, utility, durability, and comfort.

Leadership in Energy and Environmental Design (LEED) is a set of rating systems for the design, construction, operation, and maintenance of green buildings which was Developed by the U.S. Green Building Council. Other certificates system that confirms the sustainability of buildings is the British BREEAM (Building Research Establishment Environmental Assessment Method) for buildings and large scale developments. Currently, World Green Building Council is conducting research on the effects of green buildings on the health and productivity of their users and is working with World Bank to promote Green Buildings in Emerging Markets through EDGE Excellence in Design for Greater Efficiencies Market Transformation Program and certification.

Although new technologies are constantly being developed to complement current practices in creating greener structures, the common objective of green buildings is to reduce the overall impact of the built environment on human health and the natural environment by:

- Efficiently using energy, water, and other resources.
- Protecting occupant health and improving employee productivity.
- Reducing waste, pollution and environmental degradation.

A similar concept is natural building, which is usually on a smaller scale and tends to focus on the use of natural materials that are available locally. Other related topics include sustainable design and green architecture. Sustainability may be defined as meeting the needs of present generations without compromising the ability of future generations to meet their needs. Although some green building programs don't address the issue of the retrofitting existing homes, others do, especially through public schemes for energy efficient refurbishment. Green construction principles can easily be applied to retrofit work as well as new construction.

A 2009 report by the U.S. General Services Administration found 12 sustainably-designed buildings that cost less to operate and have excellent energy performance. In addition, occupants were overall more satisfied with the building than those in typical commercial buildings.These are eco-friendly buildings.

Reducing Environmental Impact

Hanging gardens of One Central Park, Sydney.

Globally buildings are responsible for a huge share of energy, electricity, water and materials consumption. The building sector has the greatest potential to deliver significant cuts in emissions at little or no cost. Buildings account for 18% of global emissions today, or the equivalent of 9 billion tonnes of CO2 annually. If new technologies in construction are not adopted during this time of rapid growth, emissions could double by 2050, according to the United Nations Environment Program.Green building practices aim to reduce the environmental impact of building. Since construction almost always degrades a building site, not building at all is preferable to green building, in terms of reducing environmental impact. The second rule is that every building should be as small as possible. The third rule is not to contribute to sprawl, even if the most energy-efficient, environmentally sound methods are used in design and construction.

Buildings account for a large amount of land. According to the National Resources Inventory, approximately 107 million acres (430,000 km^2) of land in the United States are developed. The International Energy Agency released a publication that estimated that existing buildings are responsible for more than 40% of the world's total primary energy consumption and for 24% of global carbon dioxide emissions.

Goals of Green Building

Blu Homes mkSolaire, a green building designed by Michelle Kaufmann.

The concept of sustainable development can be traced to the energy (especially fossil oil) crisis and environmental pollution concerns of the 1960s and 1970s. The Rachel Carson book, "Silent Spring", published in 1962, is considered to be one of the first initial efforts to describe sustainable development as related to green building. The green building movement in the U.S. originated from the need and desire for more energy efficient and environmentally friendly construction practices. There are a number of motives for building green, including environmental, economic, and social benefits. However, modern sustainability initiatives call for an integrated and synergistic design to both new construction and in the retrofitting of existing structures. Also known as sustainable design, this approach integrates the building life-cycle with each green practice employed with a design-purpose to create a synergy among the practices used.

Taipei 101, the tallest and largest green building of LEED Platinum certification in the world since 2011.

Green building brings together a vast array of practices, techniques, and skills to reduce and ultimately eliminate the impacts of buildings on the environment and human health. It often emphasizes taking advantage of renewable resources, e.g., using sunlight through passive solar, active solar, and photovoltaic equipment, and using plants and trees through green roofs, rain gardens, and reduction of rainwater run-off. Many other techniques are used, such as using low-impact building materials or using packed gravel or permeable concrete instead of conventional concrete or asphalt to enhance replenishment of ground water.

While the practices or technologies employed in green building are constantly evolving and may differ from region to region, fundamental principles persist from which the method is derived: siting and structure design efficiency, energy efficiency, water efficiency, materials efficiency, indoor environmental quality enhancement, operations and maintenance optimization and waste and toxics reduction. The essence of green building is an optimization of one or more of these principles. Also, with the proper synergistic design, individual green building technologies may work together to produce a greater cumulative effect.

On the aesthetic side of green architecture or sustainable design is the philosophy of designing a building that is in harmony with the natural features and resources surrounding the site. There are several key steps in designing sustainable buildings: specify 'green' building materials from local sources, reduce loads, optimize systems, and generate on-site renewable energy.

Life Cycle Assessment

A life cycle assessment (LCA) can help avoid a narrow outlook on environmental, social and economic concerns by assessing a full range of impacts associated with all cradle-to-grave stages of a process: from extraction of raw materials through materials processing, manufacture, distribution, use, repair and maintenance, and disposal or recycling. Impacts taken into account include (among others) embodied energy, global warming potential, resource use, air pollution, water pollution, and waste.

In terms of green building, the last few years have seen a shift away from a *prescriptive* approach, which assumes that certain prescribed practices are better for the environment, toward the scientific evaluation of actual performance through LCA.

Although LCA is widely recognized as the best way to evaluate the environmental impacts of buildings (ISO 14040 provides a recognized LCA methodology), it is not yet a consistent requirement of green building rating systems and codes, despite the fact that embodied energy and other life cycle impacts are critical to the design of environmentally responsible buildings.

In North America, LCA is rewarded to some extent in the Green Globes® rating system, and is part of the new American National Standard based on Green Globes, *ANSI/GBI 01-2010: Green Building Protocol for Commercial Buildings*. LCA is also included as a pilot credit in the LEED system, though a decision has not been made as to whether it will be incorporated fully into the next major revision. The state of California also included LCA as a voluntary measure in its 2010 draft *Green Building Standards Code*.

Although LCA is often perceived as overly complex and time consuming for regular use by design professionals, research organizations such as BRE in the UK and the Athena Sustainable Materials Institute in North America are working to make it more accessible.

In the UK, the BRE *Green Guide to Specifications* offers ratings for 1,500 building materials based on LCA.

In North America, the ATHENA® *EcoCalculator for Assemblies* provides LCA results for several hundred common building assembles based on data generated by its more complex parent software, the ATHENA® *Impact Estimator for Buildings*. (The EcoCalculator is available free at www.athenasmi.org.) Athena software tools are especially useful early in the design process when material choices have far-reaching implications for overall environmental impact. They allow designers to experiment with different material mixes to achieve the most effective combination.

Siting and Structure Design Efficiency

The foundation of any construction project is rooted in the concept and design stages. The concept stage, in fact, is one of the major steps in a project life cycle, as it has the largest impact on cost and performance. In designing environmentally optimal buildings, the objective is to minimize the total environmental impact associated with all life-cycle stages of the building project.

Exterior Light Shelves - Green Office Building, Denver Colorado.

However, building as a process is not as streamlined as an industrial process, and varies from one building to the other, never repeating itself identically. In addition, buildings are much more complex products, composed of a multitude of materials and components each constituting various design variables to be decided at the design stage. A variation of every design variable may affect the environment during all the building's relevant life-cycle stages.

Energy Efficiency

An eco-house at Findhorn Ecovillage with a turf roof and solar panels.

Green buildings often include measures to reduce energy consumption – both the embodied energy required to extract, process, transport and install building materials and operating energy to provide services such as heating and power for equipment.

As high-performance buildings use less operating energy, embodied energy has assumed much greater importance – and may make up as much as 30% of the overall life cycle energy consumption. Studies such as the U.S. LCI Database Project show buildings built primarily with wood will have a lower embodied energy than those built primarily with brick, concrete, or steel.

To reduce operating energy use, designers use details that reduce air leakage through the building envelope (the barrier between conditioned and unconditioned space). They also specify high-performance windows and extra insulation in walls, ceilings, and floors. Another strategy, passive solar building design, is often implemented in low-energy homes. Designers orient windows and walls and place awnings, porches, and trees to shade windows and roofs during the summer while maximizing solar gain in the winter. In addition, effective window placement (daylighting) can provide more natural light and lessen the need for electric lighting during the day. Solar water heating further reduces energy costs.

Onsite generation of renewable energy through solar power, wind power, hydro power, or biomass can significantly reduce the environmental impact of the building. Power generation is generally the most expensive feature to add to a building.

Water Efficiency

Reducing water consumption and protecting water quality are key objectives in sustainable building. One critical issue of water consumption is that in many areas, the demands on the supplying aquifer exceed its ability to replenish itself. To the maximum extent feasible, facilities should increase their dependence on water that is collected, used, purified, and reused on-site. The protection and conservation of water throughout the life of a building may be accomplished by designing for dual plumbing that recycles water in toilet flushing or by using water for washing of the cars. Waste-water may be minimized by utilizing water conserving fixtures such as ultra-low flush toilets and low-flow shower heads. Bidets help eliminate the use of toilet paper, reducing sewer traffic and increasing possibilities of re-using water on-site. Point of use water treatment and heating improves both water quality and energy efficiency while reducing the amount of water in circulation. The use of non-sewage and greywater for on-site use such as site-irrigation will minimize demands on the local aquifer.

Large commercial buildings with water and energy efficiency can qualify for an LEED Certification. Philadelphia's Comcast Center is the tallest building in Philadelphia. It's also one of the tallest buildings in the USA that is LEED Certified. Their environmental engineering consists of a hybrid central chilled water system which cools floor-by-floor with steam instead of water. Burn's Mechanical set-up the entire renovation of the 58 story, 1.4 million square foot sky scraper.

Materials Efficiency

Building materials typically considered to be 'green' include lumber from forests that have been certified to a third-party forest standard, rapidly renewable plant materials like bamboo and straw, dimension stone, recycled stone, recycled metal and other products that are non-toxic, reusable, renewable, and/or recyclable. For concrete a high performance or Romanself-healing concrete is available. The EPA (Environmental Protection Agency) also suggests using recycled industrial goods, such as coal combustion products,foundry sand, and demolition debris in construction projects. Energy efficientbuilding materials and appliances are promoted in the United States through energy rebate programs.

Indoor Environmental Quality Enhancement

The Indoor Environmental Quality (IEQ) category in LEED standards, one of the five environmental categories, was created to provide comfort, well-being, and productivity of occupants. The LEED IEQ category addresses design and construction guidelines especially: indoor air quality (IAQ), thermal quality, and lighting quality.

Indoor Air Quality seeks to reduce volatile organic compounds, or VOCs, and other air impurities such as microbial contaminants. Buildings rely on a properly designed ventilation system (passively/naturally or mechanically powered) to provide adequate ventilation of cleaner air from outdoors or recirculated, filtered air as well as isolated operations (kitchens, dry cleaners, etc.) from other occupancies. During the design and construction process choosing construction materials and interior finish products with zero or low VOC emissions will improve IAQ. Most building materials and cleaning/maintenance products emit gases, some of them toxic, such as many VOCs including formaldehyde. These gases can have a detrimental impact on occupants' health, comfort, and productivity. Avoiding these products will increase a building's IEQ. LEED, HQE and Green Star contain specifications on use of low-emitting interior. Draft LEED 2012 is about to expand the scope of the involved products. BREEAM limits formaldehyde emissions, no other VOCs. MAS Certified Green is a registered trademark to delineate low VOC-emitting products in the marketplace. The MAS Certified Green Program ensures that any potentially hazardous chemicals released from manufactured products have been thoroughly tested and meet rigorous standards established by independent toxicologists to address recognized long term health concerns. These IAQ standards have been adopted by and incorporated into the following programs: (1) The United States Green Building Council (USGBC) in their LEED rating system (2) The California Department of Public Health (CDPH) in their section 01350 standards (3) The Collaborative for High Performance Schools (CHPS) in their Best Practices Manual and (4) The Business and Institutional Furniture Manufacturers Association (BIFMA) in their level® sustainability standard.

Also important to indoor air quality is the control of moisture accumulation (dampness) leading to mold growth and the presence of bacteria and viruses as well as dust mites and other organisms and microbiological concerns. Water intrusion through a building's envelope or water condensing on cold surfaces on the building's interior can enhance and sustain microbial growth. A well-insulated and tightly sealed envelope will reduce moisture problems but adequate ventilation is also necessary to eliminate moisture from sources indoors including human metabolic processes, cooking, bathing, cleaning, and other activities.

Personal temperature and airflow control over the HVAC system coupled with a properly designed building envelope will also aid in increasing a building's thermal quality. Creating a high performance luminous environment through the careful integration of daylight and electrical light sources will improve on the lighting quality and energy performance of a structure.

Solid wood products, particularly flooring, are often specified in environments where occupants are known to have allergies to dust or other particulates. Wood itself is considered to be hypo-allergenic and its smooth surfaces prevent the buildup of particles common in soft finishes like carpet. The Asthma and Allergy Foundation of American recommends hardwood, vinyl, linoleum tile or slate flooring instead of carpet. The use of wood products can also improve air quality by absorbing or releasing moisture in the air to moderate humidity.

Interactions among all the indoor components and the occupants together form the processes that determine the indoor air quality. Extensive investigation of such processes is the subject of indoor air scientific research and is well documented in the journal Indoor Air.

Operations and Maintenance Optimization

No matter how sustainable a building may have been in its design and construction, it can only remain so if it is operated responsibly and maintained properly. Ensuring operations and maintenance(O&M) personnel are part of the project's planning and development process will help retain the green criteria designed at the onset of the project. Every aspect of green building is integrated into the O&M phase of a building's life. The addition of new green technologies also falls on the O&M staff. Although the goal of waste reduction may be applied during the design, construction and demolition phases of a building's life-cycle, it is in the O&M phase that green practices such as recycling and air quality enhancement take place. O&M staff should aim to establish best practices in energy efficiency, resource conservation, ecologically sensitive products and other sustainable practices. Education of building operators and occupants is key to effective implementation of sustainable strategies in O&M services.

Waste Reduction

Green architecture also seeks to reduce waste of energy, water and materials used during construction. For example, in California nearly 60% of the state's waste comes from commercial buildings During the construction phase, one goal should be to reduce the amount of material going to landfills. Well-designed buildings also help reduce the amount of waste generated by the occupants as well, by providing on-site solutions such as compost bins to reduce matter going to landfills.

To reduce the amount of wood that goes to landfill, Neutral Alliance (a coalition of government, NGOs and the forest industry) created the website dontwastewood.com. The site includes a variety of resources for regulators, municipalities, developers, contractors, owner/operators and individuals/homeowners looking for information on wood recycling.

When buildings reach the end of their useful life, they are typically demolished and hauled to landfills. Deconstruction is a method of harvesting what is commonly considered "waste" and reclaiming it into useful building material. Extending the useful life of a structure also reduces waste – building materials such as wood that are light and easy to work with make renovations easier.

To reduce the impact on wells or water treatment plants, several options exist. "Greywater", wastewater from sources such as dishwashing or washing machines, can be used for subsurface irrigation, or if treated, for non-potable purposes, e.g., to flush toilets and wash cars. Rainwater collectors are used for similar purposes.

Centralized wastewater treatment systems can be costly and use a lot of energy. An alternative to this process is converting waste and wastewater into fertilizer, which avoids these costs and shows other benefits. By collecting human waste at the source and running it to a semi-centralized biogas plant with other biological waste, liquid fertilizer can be produced. This concept was demonstrated by a settlement in Lubeck Germany in the late 1990s. Practices like these provide soil with organic nutrients and create carbon sinks that remove carbon dioxide from the atmosphere, offsetting greenhouse gas emission. Producing artificial fertilizer is also more costly in energy than this process.

Cost and Payoff

The most criticized issue about constructing environmentally friendly buildings is the price. Photo-voltaics, new appliances, and modern technologies tend to cost more money. Most green buildings cost a premium of <2%, but yield 10 times as much over the entire life of the building. In regards to the financial benefits of green building, "Over 20 years, the financial payback typically exceeds the additional cost of greening by a factor of 4-6 times. And broader benefits, such as reductions in greenhouse gases

(GHGs) and other pollutants have large positive impacts on surrounding communities and on the planet." The stigma is between the knowledge of up-front cost vs. life-cycle cost. The savings in money come from more efficient use of utilities which result in decreased energy bills. It is projected that different sectors could save $130 Billion on energy bills. Also, higher worker or student productivity can be factored into savings and cost deductions.

Numerous studies have shown the measurable benefit of green building initiatives on worker productivity. In general it has been found that, "there is a direct correlation between increased productivity and employees who love being in their work space." Specifically, worker productivity can be significantly impacted by certain aspects of green building design such as improved lighting, reduction of pollutants, advanced ventilation systems and the use of non-toxic building materials. In "The Business Case for Green Building", the U.S. Green Building Council gives another specific example of how commercial energy retrofits increase worker health and thus productivity, "People in the U.S. spend about 90% of their time indoors. EPA studies indicate indoor levels of pollutants may be up to ten times higher than outdoor levels. LEED-certified buildings are designed to have healthier, cleaner indoor environmental quality, which means health benefits for occupants."

Studies have shown over a 20-year life period, some green buildings have yielded $53 to $71 per square foot back on investment. Confirming the rentability of green building investments, further studies of the commercial real estate market have found that LEED and Energy Star certified buildings achieve significantly higher rents, sale prices and occupancy rates as well as lower capitalization rates potentially reflecting lower investment risk.

Regulation and Operation

As a result of the increased interest in green building concepts and practices, a number of organizations have developed standards, codes and rating systems that let government regulators, building professionals and consumers embrace green building with confidence. In some cases, codes are written so local governments can adopt them as bylaws to reduce the local environmental impact of buildings.

Green building rating systems such as BREEAM (United Kingdom), LEED (United States and Canada), DGNB (Germany), CASBEE (Japan), and VERDE^{GBCe} (Spain) help consumers determine a structure's level of environmental performance. They award credits for optional building features that support green design in categories such as location and maintenance of building site, conservation of water, energy, and building materials, and occupant comfort and health. The number of credits generally determines the level of achievement.

Green building codes and standards, such as the International Code Council's draft International Green Construction Code, are sets of rules created by standards development organizations that establish minimum requirements for elements of green building such as materials or heating and cooling.

Some of the major building environmental assessment tools currently in use include:

- United States: International Green Construction Code (IGCC)

International Frameworks and Assessment Tools

IPCC Fourth Assessment Report

Climate Change 2007, the Fourth Assessment Report (AR4) of the United Nations Intergovernmental Panel on Climate Change (IPCC), is the fourth in a series of such reports. The IPCC was established by the World Meteorological Organization (WMO) and the United Nations Environment Programme (UNEP) to assess scientific, technical and socio-economic information concerning climate change, its potential effects and options for adaptation and mitigation.

UNEP and Climate Change

United Nations Environment Program UNEP works to facilitate the transition to low-carbon societies, support climate proofing efforts, improve understanding of climate change science, and raise public awareness about this global challenge.

GHG Indicator

The Greenhouse Gas Indicator: UNEP Guidelines for Calculating Greenhouse Gas Emissions for Businesses and Non-Commercial Organizations.

Agenda 21

Agenda 21 is a programme run by the United Nations (UN) related to sustainable development. It is a comprehensive blueprint of action to be taken globally, nationally and locally by organizations of the UN, governments, and major groups in every area in which humans impact on the environment. The number 21 refers to the 21st century.

Fidic's PSM

The International Federation of Consulting Engineers (FIDIC) Project Sustainability Management Guidelines were created in order to assist project engineers and other stakeholders in setting sustainable development goals for their projects that are recognized and accepted by as being in the interests of society as a whole. The process is also intended to allow the alignment of project goals with local conditions and priorities and to assist those involved in managing projects to measure and verify their progress.

The Project Sustainability Management Guidelines are structured with Themes and Sub-Themes under the three main sustainability headings of Social, Environmental and Economic. For each individual Sub-Theme a core project indicator is defined along with guidance as to the relevance of that issue in the context of an individual project.

The Sustainability Reporting Framework provides guidance for organizations to use as the basis for disclosure about their sustainability performance, and also provides stakeholders a universally applicable, comparable framework in which to understand disclosed information.

The Reporting Framework contains the core product of the Sustainability Reporting Guidelines, as well as Protocols and Sector Supplements. The Guidelines are used as the basis for all reporting. They are the foundation upon which all other reporting guidance is based, and outline core content for reporting that is broadly relevant to all organizations regardless of size, sector, or location. The Guidelines contain principles and guidance as well as standard disclosures – including indicators – to outline a disclosure framework that organizations can voluntarily, flexibly, and incrementally, adopt.

Protocols underpin each indicator in the Guidelines and include definitions for key terms in the indicator, compilation methodologies, intended scope of the indicator, and other technical references.

Sector Supplements respond to the limits of a one-size-fits-all approach. Sector Supplements complement the use of the core Guidelines by capturing the unique set of sustainability issues faced by different sectors such as mining, automotive, banking, public agencies and others.

IPD Environment Code

The IPD Environment Code was launched in February 2008. The Code is intended as a good practice global standard for measuring the environmental performance of corporate buildings. Its aim is to accurately measure and manage the environmental impacts of corporate buildings and enable property executives to generate high quality, comparable performance information about their buildings anywhere in the world. The Code covers a wide range of building types (from offices to airports) and aims to inform and support the following;

- Creating an environmental strategy.
- Inputting to real estate strategy.
- Communicating a commitment to environmental improvement.
- Creating performance targets.
- Environmental improvement plans.

- Performance assessment and measurement.
- Life cycle assessments.
- Acquisition and disposal of buildings.
- Supplier management.
- Information systems and data population.
- Compliance with regulations.
- Team and personal objectives.

IPD estimate that it will take approximately three years to gather significant data to develop a robust set of baseline data that could be used across a typical corporate estate.

ISO 21931

ISO/TS 21931:2006, Sustainability in building construction—Framework for methods of assessment for environmental performance of construction works—Part 1: Buildings, is intended to provide a general framework for improving the quality and comparability of methods for assessing the environmental performance of buildings. It identifies and describes issues to be taken into account when using methods for the assessment of environmental performance for new or existing building properties in the design, construction, operation, refurbishment and deconstruction stages. It is not an assessment system in itself but is intended be used in conjunction with, and following the principles set out in, the ISO 14000 series of standards.

Groundwater Banking

Groundwater banking is a water management tool designed to increase water supply reliability. Groundwater can be created by using dewatered aquifer space to store water during the years when there is abundant rainfall. It can then be pumped and used during years that don't have a surplus of water. People can manage the use of groundwater to benefit society through the purchasing and selling of these groundwater rights. The surface water should be used first, and then the groundwater will be used when there isn't enough surface water to meet demands. The groundwater will reduce the risk of relying on surface water and will maximize expected income. There are regulatory storage-type aquifer recovery and storage systems which when water is injected into it gives the right to withdraw the water later on. Groundwater banking has been implemented into semiarid and arid southwestern United States because this is where there is the most need for extra water. The overall goal is to transfer water from low-value to high-value uses by bringing buyers and sellers together.

Groundwater Storage Concepts

The bank is an aquifer used as an underground storage tank, and the recharge of water causes an increase in the volume of water stored in the aquifer to have increasing water levels. In the case of a withdrawal there would be a decrease in water levels. The amount of water depends also on a couple of other factors including groundwater pumping by other users, leakage, and natural recharge. The recharge of water by land application or injection increases the volume of water, and then some of the water will be used at a future time. It can be looked at as inputs of water minus outputs is equal to the change in water storage.

Another aspect is the hydrology which is the difference between dynamic and static response to recharge and abstraction. The water levels will rise or fall in the well during recharge and recovery. Once recharge and recovery stops the water levels return to background levels, and one of the main issues is the change in static water levels after the dynamic response from recharge or recovery disappears.

There can be some technical issues with the aquifer response to manage recharge and recovery. If the aquifer is hydraulically connected to body of water on the surface there are increases in the water table elevation as the result of managed recharge. This could increase the rate of discharge or decrease the amount of induced recharge, and both of these cause water to leave the basin. The recharge and recovery could also affect the lateral and vertical groundwater flow into the aquifer. There isn't always a one-to-one correspondence between the volume of water and the change in storage.

Methods of Groundwater Banking

“'Groundwater banking”' is accomplished in two ways: through in-lieu and direct recharge. In-lieu recharge is storing water by utilizing surface water “in-lieu” of pumping groundwater, thereby storing an equal amount in the groundwater basin. In-lieu recharge is the renewable surface water used to irrigate the farmland in place of using regular groundwater. This is helping to save more groundwater because the water stays in the aquifer to be used later. Direct recharge is storing water by allowing it to percolate directly to storage in the groundwater basin. With direct recharge it floods an area so that water seeps through the ground to get to the aquifers. The water is then pumped out when there is more of a demand with the use of recovery wells.

Pros and Cons

There are some disadvantages to retrieving this groundwater. As groundwater is withdrawn from below the Earth's surface, the ground above settles. This settling of land, known as subsidence, can fracture roads and building foundations and can burst water, sewer, and gas lines. This method hasn't been well tested yet, so there could be some negative impacts on the environment. Water is a public resource and this could make water become a private industry.

Groundwater banking can be compared and contrasted to the use of surface reservoirs. Groundwater banking has many advantages over the use of surface reservoirs. The projects do not cost as much to construct and store the same amount, if not more, than the surface reservoirs. The bank will have less of an impact on the environment than a surface reservoir. The water that is in the bank will no longer be exposed to evaporation, but several feet per year of water is lost in the reservoirs. It is more reliable to use when the climate is changing and can respond to seasonal changes better to manage the water than the surface reservoirs.

There are also some disadvantages to groundwater banking over surface reservoirs. There are energy costs to recovering the water and these costs are usually more than the reservoirs. There is also a pumping capacity and when the demands change during the year the productiveness can be limited.

Not all groundwater is used when sold. Some groundwater is being studied for its benefits. Groundwater banking and aquifer storage systems are being explored to control flooding during times of high precipitation.

The groundwater is being traded in many regions. There are trades even in the United States. The city of San Antonio, Texas is the largest city in the United States that relies solely on groundwater for its municipal supply.

Feasibility of Groundwater Recharge on Agricultural Land

There haven't been many successful trials of groundwater banking on agricultural land since the land is usually privately owned. The owners have to be on board with the practice of groundwater banking knowing what the risks and best practices entail. A study was done to find a Soil Agricultural Groundwater Banking Index (SAGBI) which evaluates soil suitability for the use of groundwater banking in California. There are five factors that determine the feasibility of groundwater recharge on agricultural land: deep percolation, root zone residence time, topography, chemical limitations, and soil surface conditions. The five factors were modeled using United States Department of Agriculture Natural Resources Conservation Service (USDA-NRCS) digital soil survey data.

For the deep percolation factor a high rate of water transmission through the soil profile and into the aquifer below is the key to successful groundwater banking. It becomes more important when there is flooding since it could be used as the main water source. It is derived from the saturated hydraulic conductivity of the limiting layer. Saturated hydraulic conductivity measures soil permeability when the soil is saturated.

When looking at root zone residence time factor it was found that a prolonged duration of saturated conditions in the root zone has the possibility to cause damage to perennial crops. If the soil causes a bud break it is more likely that the crop will become damaged. Most crops aren't able to withstand long periods of saturated conditions in the root zone. The root zone residence time estimates the likelihood of having good enough drainage within the root zone once water is applied.

The topography of land for spreading water across fields has the best outcome when there is level topography. Level topography works the best because it holds water better on the landscape which allows infiltration across large areas. Infiltration reduces ponding and minimizes erosion by runoff.

The chemical limitations factor is related to the salinity which is a threat to the sustainability of agriculture and groundwater. This factor was determined by electrical conductivity (EC) of the soil which measures the soil salinity. The best soil has the lowest levels of salinity. Soil also has pesticides and nitrate, but it is unable to be evaluated due to the dependency on management history.

The surface condition factor is when banking by flood spreading can change the soil surfaces physical conditions. Infiltrations limits can be caused by quality and depth of water that could lead to the destruction of aggregates, the formation of physical crusts, and compaction. To determine soil condition two factors were examined: soil erosion factor and sodium absorption ratio (SAR).

To determine the feasibility of groundwater banking each of the five factors were assigned a weight to how significant it was, and then a SAGBI score was calculated. The weights were 27.5% deep percolation, 27.5% root zone residence time, 20% topography, 20% chemical limitations, and 5% surface conditions. Of the 17.5 million acres of agricultural land examined only 5 million acres were considered soils with excellent, good, and moderately good suitability.

Agricultural groundwater banking can be associated with financial risk which may cause crop loss, so in the end the loss may exceed the benefits of water saving. Adoption of this practice would require support to protect growers from risk of crop failure.

Groundwater Accounting System

The accounting system tracks the recharge and withdrawals of stored water and it can include a market system to reserve the storage of water. Depositors could earn credits for the recharge of water which can be used later on for the water recovered from the bank. The banking system could then be set up to allow trading of credits. There are several objectives of the water bank accounting system: track water deposits, withdrawals, and to control the amount, timing, and location of withdrawals by the participants. If groundwater isn't regulated there is more of a chance for freeriding and overuse. There needs to be sustainability of the system in order to continue with the operation of the system or there would be no point to using it.

The accounting method that will be used is the double-entry accounting method, so every transaction is recorded as a debit and a credit in separate ledger accounts. This also allows for tracking of inventory in asset accounts and claims to inventory in ownership accounts. Deposits happen when more water is stored in an aquifer than there is supposed to be. The recharge of water by a member would be a credit in a member's account and a liability in the bank's account. For the bank to be successful then both ledgers have to be balanced, so the right to water in a member's accounts should be equal to the amount of water that can be recovered from the system. If the right to water is greater than liabilities then the bank is insolvent, and this will become a problem when a drought occurs.

There are some issues that could arise from using this water bank accounting system. These problems were evaluated for the Las Posas Basin groundwater bank and Fox Canyon Groundwater Management Agency (FCGMA) has jurisdiction over the project. FCGMA reported that the accumulation of credits has been increasing for banks. What can happen is the accumulated credits can become greater than the annual abstraction rate. The volume of credits accumulating exceed the amount of water that can be taken out during a short-time period. This will cause a threat to the regional groundwater resource or even depressions in groundwater elevations, land subsidence, and seawater intrusion.

Regulatory Framework

The banking systems need regulatory control over the basin to implement the withdrawal rates and to ensure that other participants won't abstract too much stored water. The best scenario would be that the bank owner or participants would be the main users to ensure that abstractions are controlled. If this is not the case, then there must be another way to control the number and amount of abstractions happening. It needs to be clear who has priority over stored water, so that when abstractions are constrained it is known who will get the water first. There can be problems when multiple entities have jurisdiction over a project, and this can cause regulatory and organizational challenges. There are some generally accepted rules and also many of the issues are handled through state-specific concepts. One of the main requirements is an action needs to be in place so that the stored water isn't being abstracted by other users who aren't involved in the system. These frameworks rely on the knowledge of hydrogeology to determine the success of a system, and the systems need to provide benefits to prove it was worth building.

Economics

The different projects can become economically efficient by maximizing the benefits of the limited resource (water). To maximize efficiency the users need to find where marginal cost is equal to marginal benefit. It is important for supply to equal demand like in the figure below. The use of water becomes a negative externality when there is rivalry and the property rights are not well-defined. The way to eliminate some of the negative externality is there can be a tax placed on the resource to increase the marginal cost. When they tax the right amount the user will use the resource at the socially acceptable level. The other way to affect the externality is to create a subsidy. A subsidy will increase the marginal benefit in order to get to the socially accepted level of use for the resource.

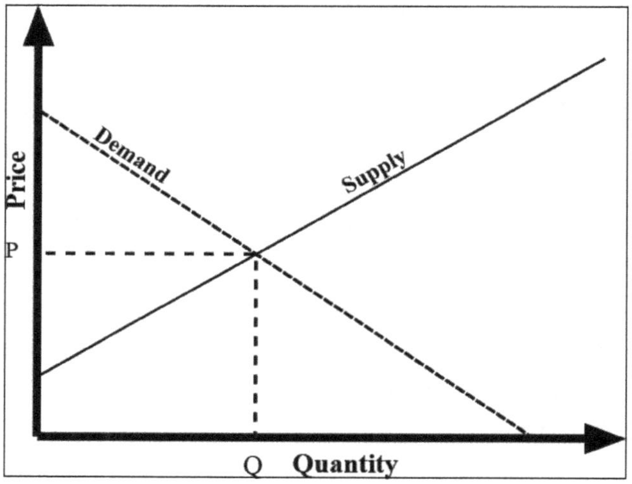

Water is not a homogeneous commodity for several reasons which include sensitivity to location, time of use, form of the water, and administrative responses. The use of groundwater banking can make water a more homogenous commodity. This can create a market value which will enhance private investment increasing the benefits. It will also align marginal benefit with marginal cost causing the market to come to a economically-efficient level.

Water has high transaction costs and create market barriers which devalues the use to society restricting the reallocation of resources. The demand does not change when there is a market barrier so there will be many unpleasant people if they don't get their share of water. If a bank is in the process of being made the removal of market-access barriers can be part of the negotiations, but it isn't necessarily the bank itself that is the cause. Groundwater banking could reduce transaction costs because each individual won't have to analyze each transaction.

Permeable Paving

Permeable paving is a range of sustainable materials and techniques for permeable pavements with a base and subbase that allow the movement of stormwater through the surface. In addition to reducing runoff, this effectively traps suspended solids and filters pollutants from the water. Examples include roads, paths, lawns and lots that are subject to light vehicular traffic, such as car/parking lots, cycle-paths, service or emergency access lanes, road and airport shoulders, and residential sidewalks and driveways.

Permeable paving demonstration.

Stone paving in Santarém, Portugal.

Although some porous paving materials appear nearly indistinguishable from nonporous materials, their environmental effects are qualitatively different. Whether it is pervious concrete, porous asphalt, paving stones or concrete or plastic-based pavers, all these pervious materials allow stormwater to percolate and infiltrate the surface areas, traditionally impervious to the soil below. The goal is to control stormwater at the source, reduce runoff and improve water quality by filtering pollutants in the substrata layers.

Description and Applications

Permeable solutions can be based on: porous asphalt and concrete surfaces, concrete pavers (permeable interlocking concrete paving systems - PICP), or polymer-based grass pavers, grids and geocells. Porous pavements and concrete pavers (actually the voids in-between them) enable stormwater to drain through a stone base layer for on-site infiltration and filtering. Polymer based grass grid or cellular paver systems provide load bearing reinforcement for unpaved surfaces of gravel or turf.

Grass pavers, plastic turf reinforcing grids (PTRG), and geocells (cellular confinement systems) are honeycombed 3D grid-cellular systems, made of thin-walled HDPE plastic or other polymer alloys. These provide grass reinforcement, ground stabilization and gravel retention. The 3D structure reinforces infill and transfers vertical loads from the surface, distributing them over a wider area. Selection of the type of cellular grid depends to an extent on the surface material, traffic and loads. The cellular grids are installed on a prepared base layer of open-graded stone (higher void spacing) or engineered stone (stronger). The surface layer may be compacted gravel or topsoil seeded with grass and fertilizer. In addition to load support, the cellular grid reduces compaction of the soil to maintain permeability, while the roots improve permeability due to their root channels.

In new suburban growth, porous pavements protect watersheds. In existing built-up areas and towns, redevelopment and reconstruction are opportunities to implement stormwater water management practices. Permeable paving is an important component in Low Impact Development (LID), a process for land development in the United States that attempts to minimize impacts on water quality and the similar concept of sustainable drainage systems (SuDS) in the United Kingdom.

The infiltration capacity of the native soil is a key design consideration for determining the depth of base rock for stormwater storage or for whether an underdrain system is needed.

Advantages

Managing Runoff

Permeable paving surfaces have been demonstrated as effective in managing runoff from paved surfaces. Large volumes of urban runoff causes serious erosion and siltation in surface water bodies.

Controlling Pollutants

Permeable paving surfaces keep the pollutants in place in the soil or other material underlying the roadway, and allow water seepage to groundwater recharge while preventing the stream erosion problems. They capture the heavy metals that fall on them, preventing them from washing downstream and accumulating inadvertently in the environment. In the void spaces, naturally occurring micro-organisms digest car oils, leaving little but carbon dioxide and water. Rainwater infiltration is usually less than that of an impervious pavement with a separate stormwater management facility somewhere downstream..in areas where infiltration is not possible due to unsuitable soil conditions permeable pavements are used in the attenuation mode where water is retained in the pavement and slowly released to surface water systems between storm events.

Trees

Permeable pavements may give urban trees the rooting space they need to grow to full size. A "structural-soil" pavement base combines structural aggregate with soil; a porous surface admits vital air and water to the rooting zone. This integrates healthy ecology and thriving cities, with the living tree canopy above, the city's traffic on the ground, and living tree roots below. The benefits of permeables on urban tree growth have not been conclusively demonstrated and many researchers have observed tree growth is not increased if construction practices compact materials before permeable pavements are installed.

Disadvantages

Runoff Volumes

Permeable pavements are designed to replace Effective Impervious Areas (EIAs), not to manage stormwater from other impervious surfaces on site. Use of this technique must be part of an overall on site management system for stormwater, and is not a replacement for other techniques.

Also, in a large storm event, the water table below the porous pavement can rise to a higher level preventing the precipitation from being absorbed into the ground. The additional water is stored in the open graded crushed drain rock base and remains until the subgrade can absorb the water. For clay-based soils, or other low to 'non'-draining soils, it is important to increase the depth of the crushed drain rock base to allow additional capacity for the water as it waits to be infiltrated.

The best way to prevent this problem is to understand the soil infiltration rate, and design the pavement and base depths to meet the volume of water. Or, allow for adequate rain water run off at the pavement design stage.

Pollutant Load

Highly contaminated runoff can be generated by some land uses where pollutant concentrations exceed those typically found in stormwater. These "hot spots" include commercial nurseries, recycling facilities, fueling stations, industrial storage, marinas, some outdoor loading facilities, public works yards, hazardous materials generators (if containers are exposed to rainfall), vehicle service and maintenance areas, and vehicle and equipment washing and steam cleaning facilities. Since porous pavement is an infiltration practice, it should not be applied at stormwater hot spots due to the potential for groundwater contamination. All contaminated runoff should be prevented from entering municipal storm drain systems by using best management practices (BMPs) for the specific industry or activity.

Weight and Traffic Volumes

Reference sources differ on whether low or medium traffic volumes and weights are appropriate for porous pavements. For example, around truck loading docks and areas of high commercial traffic, porous pavement is sometimes cited as being inappropriate. However, given the variability of products available, the growing number of existing installations in North America and targeted research by both manufacturers and user agencies, the range of accepted applications seems to be expanding. Some concrete paver companies have developed products specifically for industrial applications. Working examples exist at fire halls, busy retail complex parking lots, and on public and private roads, including intersections in parts of North America with quite severe winter conditions.

Siting

Permeable pavements may not be appropriate when land surrounding or draining into the pavement exceeds a 20 percent slope, where pavement is down slope from buildings or where foundations have piped drainage at their footers. The key is to ensure that drainage from other parts of a site is intercepted and dealt with separately rather than being directed onto permeable surfaces.

Climate

Cold climates may present special challenges. Road salt contains chlorides that could migrate through the porous pavement into groundwater. Snow plow blades could catch block edges and damage surfaces. Sand cannot be used for snow and ice control on perveous asphalt or concrete because it will plug the pores and reduce permeability. Infiltrating runoff may freeze below the pavement, causing frost heave, though design modifications can reduce this risk. These potential problems do not mean that porous pavement cannot be used in cold climates. Porous pavement designed to reduce frost heave has been used successfully in Norway. Furthermore, experience suggests that rapid drainage below porous surfaces increases the rate of snow melt above.

Cost

Some estimates put the cost of permeable paving at two to three times that of conventional asphalt paving. Using permeable paving, however, can reduce the cost of providing larger or more stormwater BMPs on site, and these savings should be factored into any cost analysis. In addition, the off-site environmental impact costs of not reducing on-site stormwater volumes and pollution have historically been ignored or assigned to other groups (local government parks, public works and environmental restoration budgets, fisheries losses, etc.) The City of Olympia, Washington is studying the use of pervious concrete quite closely and finding that new stormwater regulations are making it a viable alternative to storm water.

Longevity and Maintenance

Some permeable pavements require frequent maintenance because grit or gravel can block the open pores. This is commonly done by industrial vacuums that suck up all the sediment. If maintenance is not carried out on a regular basis, the porous pavements can begin to function more like impervious surfaces. With more advanced paving systems the levels of maintenance needed can be greatly decreased, elastomerically bound glass pavements requires less maintenance than regular concrete paving as the glass bound pavement has 50% more void space.

Plastic grid systems, if selected and installed correctly, are becoming more and more popular with local government maintenance personnel owing to the reduction in maintenance efforts: reduced gravel migration and weed suppression in public park settings.

Some permeable paving products are prone to damage from misuse, such as drivers who tear up patches of plastic & gravel grid systems by "joy riding" on remote parking lots at night. The damage is not difficult to repair but can look unsightly in the meantime. Grass pavers require supplemental watering in the first year to establish the vegetation, otherwise they may need to be re-seeded. Regional climate also means that most grass applications will go dormant during the dry season. While brown vegetation is only a matter of aesthetics, it can influence public support for this type of permeable paving.

Traditional permeable concrete paving bricks tend to lose their color in relatively short time which can be costly to replace or clean and is mainly due to the problem of efflorescence.

Efflorescence

Efflorescence is a hardened crystalline deposit of salts, which migrate from the center of concrete or masonry pavers to the surface to form insoluble calcium carbonates that harden on the surface. Given time, these deposits form much like how a stalactite takes shape in a cave, except in this case on a flat surface. Efflorescence usually appears white, gray or black depending on the region.

Over time efflorescence begins to negatively affect the overall appearance of masonry/concrete and may cause the surfaces to become slippery when exposed to moisture. If left unchecked, this efflorescence will harden whereby the calcium/lime deposits begin to affect the integrity of the cementatious surface by slowly eroding away the cement paste and aggregate. In some cases it will also discolor stained or coated surfaces.

Efflorescence forms more quickly in areas that are exposed to excessive amounts of moisture such as near pool decks, spas, and fountains or where irrigation runoff is present. As a result, these affected regions become very slick when wet thereby causing a significant loss of "friction coefficient". This can be of serious concern especially as a public safety issue to individuals, principals and property owners by exposing them to possible injury and increased general liability claims.

Efflorescence remover chemicals can be used to remove calcium/lime build-up without damaging the integrity of the paving surface.

Types

Installation of porous pavements is no more difficult than that of dense pavements, but has different specifications and procedures which must be strictly adhered to. Nine different families of porous paving materials present distinctive advantages and disadvantages for specific applications. Here are examples:

Pervious Concrete

Pervious concrete is widely available, can bear frequent traffic, and is universally accessible. Pervious concrete quality depends on the installer's knowledge and experience.

Plastic Grids

Plastic grids allow for a 100% porous system using structural grid systems for containing and stabilizing either gravel or turf. These grids come in a variety of shapes and sizes depending on use; from pathways to commercial parking lots. These systems have been used readily in Europe for over a decade, but are gaining popularity in North America due to requirements by government for many projects to meet LEED environmental building standards. Plastic grid system are also popular with homeowners due to their lower cost to install, ease of installation, and versatility. The ideal design for this type of grid system is a closed cell system, which prevents gravel/sand/turf from migrating laterally. It is also known as Grass pavers / Turf Pavers in India.

Porous Asphalt

Porous asphalt is produced and placed using the same methods as conventional asphalt concrete; it differs in that fine (small) aggregates are omitted from the asphalt mixture. The remaining large, single-sized aggregate particles leave open voids that give the material its porosity and permeability. To ensure pavement strength, fiber may be added to the mix or a polymer-modified asphalt binder may be used. Generally, porous asphalt pavements are designed with a subsurface reservoir that holds water that passes through the pavement, allowing it to evaporate and/or percolate slowly into the surround soils.

Open-graded friction courses (OGFC) are a porous asphalt surface course used on highways to improve driving safety by removing water from the surface. Unlike a full-depth porous asphalt pavement, OGFCs do not drain water to the base of a pavement. Instead, they allow water to infiltrate the top 3/4 to 1.5 inch of the pavement and then drain out to the side of the roadway. This can improve the friction characteristics of the road and reducing road spray.

Single-Sized Aggregate

Single-sized aggregate without any binder, e.g. loose gravel, stone-chippings, is another alternative. Although it can only be safely used in very low-speed, low-traffic settings, e.g. car-parks and drives, its potential cumulative area is great.

Porous Turf

Grass pavement.

Porous turf, if properly constructed, can be used for occasional parking like that at churches and stadia. Plastic turf reinforcing grids can be used to support the increased load. Living turf transpires water, actively counteracting the "heat island" with what appears to be a green open lawn.

Permeable Interlocking Concrete Pavements

Permeable interlocking concrete pavements are concrete (or stone) units with open, permeable spaces between the units. They give an architectural appearance, and can bear both light and heavy traffic, particularly interlocking concrete pavers, excepting high-volume or high-speed roads. Some products are polymer-coated and have an entirely porous face.

Permeable Clay Brick Pavements

Permeable clay brick pavements are fired clay brick units with open, permeable spaces between the units. Clay pavers provide a durable surface that allows stormwater runoff to permeate through the joints.

Resin Bound Paving

Resin bound paving is a mixture of resin binder and aggregate. Clear resin is used to fully coat each aggregate particle before laying. Enough resin is used to allow each aggregate particle to adhere to one another and to the base yet leave voids for water to permeate through. Resin bound paving provides a strong and durable surface that is suitable for pedestrian and vehicular traffic in applications such as pathways, driveways, car parks and access roads.

Bound Recycled Glass Porous Pavement

Elastomerically bound recycled glass porous pavement consisting of bonding processed post consumer glass with a mixture of resins, pigments, granite and binding agents. Approximately 75 percent of glass in the U.S. is disposed in landfills.

Seawall

An example of a modern seawall in Ventnor on the Isle of Wight.

People socializing and walking at the Malecón, Havana.

Seawall at Urangan, Queensland.

A seawall (or sea wall) is a form of coastal defense constructed where the sea, and associated coastal processes, impact directly upon the landforms of the coast. The purpose of a seawall is to protect areas of human habitation, conservation and leisure activities from the action of tides and waves. As a seawall is a static feature it will conflict with the dynamic nature of the coast and impede the exchange of sediment between land and sea.

The coast is generally a high-energy, dynamic environment with spatial variations occurring over a wide range of temporal scales. The shoreline is part of the coastal interface which is exposed to a wide range of erosional processes arising from fluvial, aeolian and terrestrial sources, meaning that a combination of denudational processes will work against a seawall. Given the natural forces to which seawalls are constantly subjected, maintenance (and eventually replacement) is an ongoing requirement if they are to provide an effective long-term solution.

The many types in use today reflect both the varying physical forces they are designed to withstand, and location specific aspects, such as: local climate, coastal position, wave regime, and value of landform. Seawalls are classified as a hard engineering shore based structure used to provide protection and to lessen coastal erosion. However, a range of environmental problems and issues may arise from the construction of a seawall, including disrupting sediment movement and transport patterns, which are discussed in more detail below. Combined with a high construction cost, this has led to an increasing use of other soft engineering coastal management options such as beach replenishment.

Seawalls may be constructed from a variety of materials, most commonly: reinforced concrete, boulders, steel, or gabions. Additional seawall construction materials may include: vinyl, wood, aluminium, fibreglass composite, and with large biodegrable sandbags made of jute and coir. In the UK, *sea wall* also refers to an earthen bank used to create a polder, or a dike construction.

Types

A seawall works by reflecting incident wave energy back into the sea, thereby reducing the energy and erosion which the coastline would otherwise be subjected to. In addition to their unsightly visual appearance, two specific weaknesses of seawalls exist. Firstly, wave reflection induced by the wall may result in scour and subsequent lowering of the sand level of the fronting beach. Secondly, seawalls may accelerate erosion of adjacent, unprotected coastal areas because they affect the littoral drift process. The design and type of seawall that is appropriate depends on aspects specific to the location, including the surrounding erosion processes. There are three main types of seawalls: vertical, curved or stepped, and mounds, as set out in the table.

<table>
<tr><th colspan="5">Seawall types</th></tr>
<tr><th>Type</th><th>Illustration</th><th>Advantages</th><th>Disadvantages</th><th>Example</th></tr>
<tr><td rowspan="2">Vertical</td><td colspan="4">Vertical seawalls are built in particularly exposed situations. These reflect wave energy. Under storm conditions a non-breaking standing wave pattern can form, resulting in a stationary clapotic wave which moves up and down but does not travel horizontally. These waves promote erosion at the toe of the wall and can cause severe damage to the sea wall. In some cases piles are placed in front of the wall to lessen wave energy slightly.</td></tr>
<tr><td>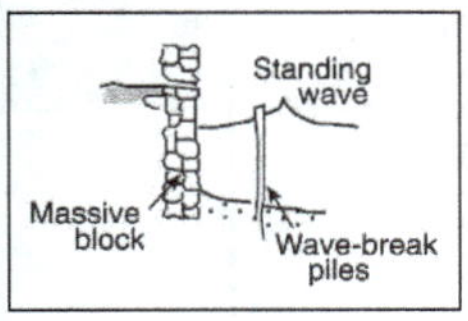
</td><td>• The first implemented, most easily designed and constructed type of seawall.
• Vertical sea walls deflect wave energy away from the coast.
• Loose rubble can absorb wave energy.</td><td>• These are partial to a lot of expensive damage in a short period of time.
• Vertical design can be undercut by high-wave energy environments over a long period of time.</td><td></td></tr>
<tr><td rowspan="2">Curved</td><td colspan="4">Curved or stepped seawalls are designed to enable waves to break to dissipate wave energy and to repel waves back to the sea. The curve can also prevent the wave overtopping the wall and provides additional protection for the toe of the wall.</td></tr>
<tr><td>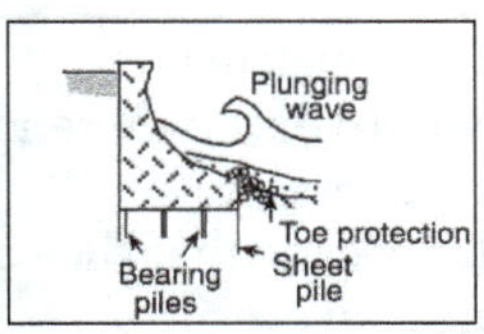
</td><td>• Concave structure introduces a dissipative element.
• The curve can prevent waves from overtopping the wall and provides extra protection for the toe of the wall
• Curved seawalls aim to re-direct most of the incident energy, resulting in low reflected waves and much reduced turbulence.</td><td>• More complex engineering and design process.
• The deflected waves can scour material at the base of the wall causing them to become undermined.</td><td></td></tr>
</table>

Mound	Mound type seawalls, using revetments or riprap, are used in less demanding settings where lower energy erosional processes operate. The least exposed sites involve the lowest-cost bulkheads and revetments of sand bags or geotextiles. These serve to armour the shore and minimise erosion and may be either watertight or porous, which allows water to filter through after the wave energy has been dissipated.			
	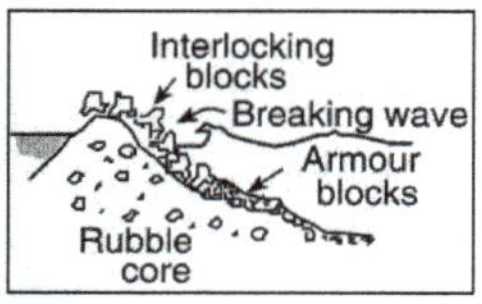	• Current designs use porous designs of rock, concrete armour. • Slope and loose material ensure maximum dissipation of wave energy. • Lower cost option.	• Less durable. • Shorter life expectancy. • Cannot withstand or protect from high-energy conditions effectively.	

Trade-Offs

A cost benefit approach is an effective way to determine whether a seawall is appropriate and whether the benefits are worth the expense. Besides controlling erosion, consideration must be given to the effects of hardening a shoreline on natural coastal ecosystems and human property or activities. A seawall is a static feature which can conflict with the dynamic nature of the coast and impede the exchange of sediment between land and sea. The table below summarises some positive and negative effects of seawalls which can be used when comparing their effectiveness with other coastal management options, such as beach nourishment.

Advantages and disadvantages of seawalls according to Short (1999)	
Advantages	**Disadvantages**
• Long term solution in comparison to soft beach nourishment. • Effectively minimizes loss of life in extreme events and damage to property caused by erosion. • Can exist longer in high energy environments in comparison to 'soft' engineering methods. • Can be used for recreation and sightseeing. • Forms a hard and strong coastal defense.	• Very expensive to construct. • Can cause beaches to dissipate rendering them useless for beach goers. • Scars the very landscape that they are trying to save and provides an 'eyesore.' • Reflected energy of waves leading to scour at base. • Can disrupt natural shoreline processes and destroy shoreline habitats such as wetlands and intertidal beaches. • Altered sediment transport processes can disrupt sand movement that can lead to increased erosion down drift from the structure.

Generally seawalls can be a successful way to control coastal erosion, but only if they are constructed well and out of materials which can withstand the force of ongoing wave energy. Some understanding is needed of the coastal processes and morphodynamics specific to the seawall location. Seawalls can be very helpful; they can offer a more long-term solution than soft engineering options, additionally providing recreation opportunities and protection from extreme events as well as everyday erosion. Extreme natural events expose weaknesses in the performance of seawalls, and analyses of these can lead to future improvements and reassessment.

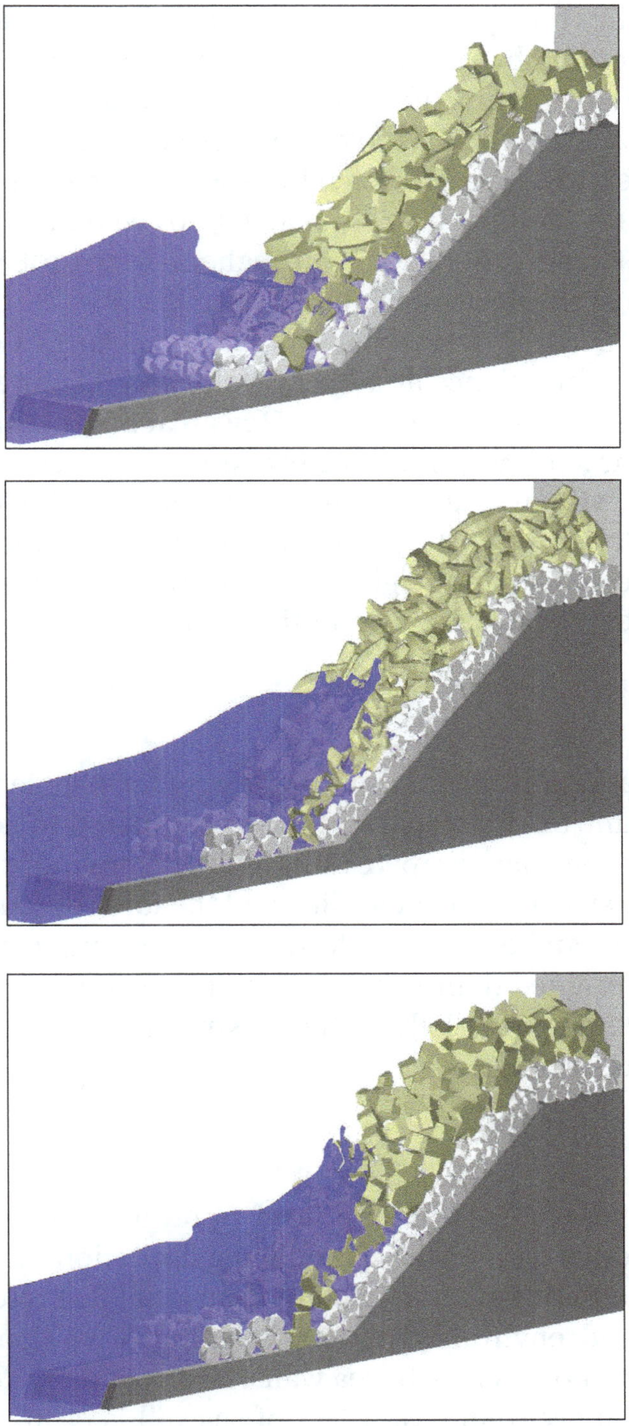

3D simulation of wave motion near a sea wall.

Issues

Sea Level Rise

Sea level rise creates an issue for seawalls worldwide as it raises both the mean normal water level and the height of waves during extreme weather events, which the current seawall heights may be unable to cope with (Allan *et al.* 1999). The most recent analyses of long, good-quality tide gauge records (corrected for GIA and when possible for other vertical land motions by the Global Positioning System, GPS) indicate a mean rate of sea level rise of 1.6–1.8 mm/yr over the twentieth century. The Intergovernmental Panel on Climate Change (IPCC) (1997) suggested that sea level rise over the next 50 – 100 years will accelerate with a projected increase in global mean sea level of +18 cm by 2050 AD. This data is reinforced by Hannah (1990) who calculated similar statistics including a rise of between +16-19.3 cm throughout 1900–1988. Super storm Sandy of 2012 is an example of the devastating effects rising sea levels can cause when mixed with a perfect storm. According to Nabil Ismail, Professor of Coastal Engineering at the Maritime Academy for Science and Technology and S. Jeffress Williams, Dept. of Geology and Geophysics at The University of Hawaii at Manoa stated that Super storm Sandy sent a storm surge of 4–5 m onto New Jersey's and New York's barrier island and urban shorelines, estimated at $70 billion in damage. EGU General Assembly Conference Abstracts. Vol. 15. 2013.This problem could be overcome by further modelling and determining the extension of height and reinforcement of current seawalls which needs to occur for safety to be ensured in both situations.

Extreme Events

Extreme events also pose a problem as it is not easy for people to predict or imagine the strength of hurricane or storm induced waves compared to normal, expected wave patterns. An extreme event can dissipate hundreds of times more energy than everyday waves, and calculating structures which will stand the force of coastal storms is difficult and, often the outcome can become unaffordable. For example, Omaha Beach seawall in New Zealand was designed to prevent erosion from everyday waves only, and when a storm in 1976 carved out 10 m behind the existing seawall the whole structure was destroyed (GeoResources, 2001).

Other Issues

Some further issues include: lack of long term trend data of seawall effects due to a relatively short duration of data records; modelling limitations and comparisons of different projects and their effects being invalid or unequal due to different beach types; materials; currents; and environments (Christchurch City Council, 2009). Lack of maintenance is also a major issue with sea walls. In a post on October 21, 2013 by NBC reporter Lucas Seiler, more than 5,000 feet of seawall is crumbling in Punta Gorda, Florida. Residents of the area pay hundreds of dollars each year into a seawall repair program. The problem is that most of the seawalls are over a half century old and are being destroyed by only heavy downpours. If not kept in check, seawalls lose effectiveness and become expensive to repair.

Historical Examples

Seawall construction has existed since ancient times. In the 1st century BCE, Romans built a seawall / breakwater at Caesarea Maritima creating an artificial harbor (Sebastos Harbor). The construction used Pozzolana concrete which hardens in contact with sea water. Barges were constructed and filled with the concrete. They were floated into position and sunk. The resulting harbor / breakwater / sea wall is still in existence today - more than 2000 years later.

More recently, sea walls were constructed in 1623 in Canvey Island, UK, when great floods of the Thames estuary occurred, prompting the construction of protection for further events in this flood prone area (Council of Europe, 1999). Since then, seawall design has become more complex and intricate in response to an improvement in materials, technology and an understanding of how coastal processes operate. This section will outline some key case studies of seawalls in chronological order and describe how they have performed in response to tsunami or ongoing natural processes and how effective they were in these situations. Analysing the successes and shortcomings of seawalls during severe natural events allows their weaknesses to be exposed, and areas become visible for future improvement.

Pondicherry

On December 26, 2004, towering waves of the 2004 Indian Ocean earthquake tsunami crashed against India's south-eastern coastline killing thousands. However, the former French colonial enclave of Pondicherry escaped unscathed. This was primarily due to French engineers who had constructed (and maintained) a massive stone seawall during the time when the city was a French colony. This 300-year-old seawall effectively kept Pondicherry's historic center dry even though tsunami waves drove water 24 ft (7.3 m) above the normal high-tide mark.

The barrier was initially completed in 1735 and over the years, the French continued to fortify the wall, piling huge boulders along its 1.25 mi (2 km) coastline to stop erosion from the waves pounding the harbor. At its highest, the barrier running along the water's edge reaches about 27 ft (8.2 m) above sea level. The boulders, some weighing up to a ton, are weathered black and brown. The sea wall is inspected every year and whenever gaps appear or the stones sink into the sand, the government adds more boulders to keep it strong (Allsop, 2002).

The Union Territory of Pondicherry recorded some 600 deaths from the huge tsunami waves that struck India's coast after the mammoth underwater earthquake (which measured 9.0 on the moment magnitude scale) off Indonesia, but most of those killed were fishermen who lived in villages beyond the artificial barrier which reinforces the effectiveness of seawalls.

Vancouver

The Vancouver Seawall is a stone seawall constructed around the perimeter of Stanley Park in Vancouver. The seawall was constructed initially as waves created by ships passing through the First Narrows were eroding the area between Prospect Point and Brockton Point. The Vancouver Seawall also exemplifies how seawalls can be utilised and valued for recreational activities and coastal sightseeing. A pedestrian, cycling and roller blading pathway exists on the seawall and has been extended far outside the parameters of Stanley Park. Construction of the seawall began in 1917, and since then this pathway has become one of the most used features of the park by both locals and tourists and now extends 22 km in total (Belyea & Ross, 1992). The construction of the seawall also provided employment for relief workers during the Great Depression and seamen from HMCS *Discovery* on Deadman's Island who were facing punishment detail in the 1950s (Steele, 1985).

Overall, the Vancouver Seawall is a prime example of how seawalls can simultaneously provide shoreline protection and a source of recreation which enhances human enjoyment of the coastal environment. It also illustrates that although shoreline erosion is a natural process, human activities, interactions with the coast and poorly planned shoreline development projects can accelerate natural erosion rates.

Japan

At least 43 percent of Japan's 29,751 km (18,486 mi) coastline is lined with concrete seawalls or other structures designed to protect the country against high waves, typhoons or even tsunamis. During the 2011 Tōhoku earthquake and tsunami, the seawalls in most areas were overwhelmed. In Kamaishi, 4-metre (13 ft) waves surmounted the seawall — the world's largest, erected a few years ago in the city's harbour at a depth of 63 m (207 ft), a length of 2 km (1.2 mi) and a cost of $1.5 billion — and eventually submerged the city center.

The risks of dependence on seawalls was most evident in the crisis at the Fukushima Dai-ichi and Fukushima Dai-ni nuclear power plants, both located along the coast close to the earthquake zone, as the tsunami washed over walls that were supposed to protect the plants. Arguably, the additional defense provided by the seawalls presented an extra margin of time for citizens to evacuate and also stopped some of the full force of energy which would have caused the wave to climb higher in the backs of coastal valleys. In contrast, the seawalls also acted in a negative way to trap water and delay its retreat.

The failure of the world's largest seawall, which cost $1.5 billion to construct, shows that building stronger sea walls to protect larger areas would have been even less cost-effective. In the case of the ongoing crisis at the nuclear power plants, higher and stronger sea walls should have been built if power plants were to be built at that site. Fundamentally, the devastation in coastal areas and a final death toll predicted to exceed 10,000 could push Japan to redesign its seawalls or consider more effective alternative methods of coastal protection for extreme events. Such hardened coastlines can also provide a false sense of security to property owners and local residents as evident in this situation.

Seawalls along the Japanese coast have also been criticized for cutting settlements off from the sea, make beaches unusable, presenting an eyesore, disturbing wildlife, and being unnecessary.

Rain Garden

Business parking lot that drains to a rain garden. The curb retains the asphalt pavement, yet lets water flow off the edges.

A rain garden is a planted depression or a hole that allows rainwater runoff from impervious urban areas, like roofs, driveways, walkways, parking lots, and compacted lawn areas, the opportunity to be absorbed. This reduces rain runoff by allowing stormwater to soak into the ground (as opposed to flowing into storm drains and surface waters which causes erosion, water pollution, flooding, and diminished groundwater). They should be designed for specific soils and climates. The purpose of a rain garden is to improve water quality in nearby bodies of water and to ensure that rainwater becomes available for plants as groundwater rather than being sent through stormwater drains straight out to sea. Rain gardens can cut down on the amount of pollution reaching creeks and streams by up to 30%.

Native and adapted plants are recommended for rain gardens because they are more tolerant of one's local climate, soil, and water conditions; have deep and variable root systems for enhanced water infiltration and drought tolerance; habitat value and diversity for local ecological communities; and overall sustainability once established. There can be trade-offs associated with using native plants, including lack of availability for some species, late spring emergence, short blooming season, and relatively slow establishment. The plants — a selection of wetland edge vegetation, such as wildflowers, sedges, rushes, ferns, shrubs and small trees — take up excess water flowing into the rain garden. Water filters through soil layers before entering the groundwater system. Root systems enhance infiltration, maintain or even augment soil permeability, provide moisture redistribution, and sustain diverse microbial populations involved in biofiltration. Also, through the process of transpiration, rain garden plants return water vapor to the atmosphere. A more wide-ranging definition covers all the possible elements that can be used to capture, channel, divert, and make the most of the natural rain and snow that falls on a property. The whole garden can become a rain garden, and each component of the whole can become a small-scale rain garden in itself.

Restoring the Water Cycle and Mitigating Urbanization

In developed areas, natural depressions where storm water would pool, are filled in. The surface of the ground is often leveled or paved. Storm water is directed into storm drains which often may cause overflows of combined sewer systems or poisoning, erosion or flooding of waterways receiving the storm water runoff. Redirected storm water is often warmer than the groundwater normally feeding a stream, and has been linked to upset in some aquatic ecosystems primarily through the reduction of dissolved oxygen (DO). Storm water runoff is also a source of a wide variety of pollutants washed off hard or compacted surfaces during rain events. These pollutants include volatile organic compounds, pesticides, herbicides, hydrocarbons and trace metals Rain gardens are designed to capture the initial flow of storm water and reduce the accumulation of toxins flowing directly into natural waterways through ground filtration. They also reduce energy consumption. For example, "the cumulative storage capacity of these rain gardens exceeds a conventional stormwater's system's by 10 times." The National Science Foundation, the United States Environmental Protection Agency, and a number of research institutions are presently studying the impact of augmenting rain gardens with materials capable of capture or chemical reduction of the pollutants to benign compounds.

Rain garden, SUNY College of Environmental Science and Forestry, Syracuse, New York.

Rain gardens are often located near a building's roof drainpipe (with or without rainwater tanks). Most rain gardens are designed to be an endpoint of drainage with a capacity to percolate all incoming water through a series of soil or gravel layers beneath the surface plantings. A French drain may be used to direct a portion of the rainwater to an overflow location for heavier rain events. By reducing peak stormwater discharge, rain gardens extend hydraulic lag time and somewhat mimic the natural water cycle displaced by urban development and allow for groundwater recharge. While rain gardens always allow for restored groundwater recharge, and reduced stormwater volumes, they may also increase pollution unless remediation materials are included in the design of the filtration layers.

The primary challenge of rain garden design centers on calculating the types of pollutants and the acceptable loads of pollutants the rain garden's filtration system can handle during storm-water events. This challenge is specifically acute when a rain event occurs after a longer dry period. The initial storm water is often highly contaminated with the accumulated pollutants from dry periods. Rain garden designers have previously focused on finding robust native plants and encouraging adequate biofiltration, but recently have begun augmenting filtration layers with media specifically suited to chemically reduce redox of incoming pollutant streams.

Rain gardens are beneficial for many reasons: improve water quality by filtering runoff, provide localized flood control, are aesthetically pleasing, and provide interesting planting opportunities. They also encourage wildlife and biodiversity, tie together buildings and their surrounding environments in attractive and environmentally advantageous ways, and provide significant partial solutions to important environmental problems that affect us all.

A rain garden provides a way to use and optimize any rain that falls, reducing or avoiding the need for irrigation. They allow a household or building to deal with excessive rainwater runoff without burdening the public storm water systems. Rain gardens differ from retention basins, in that the water will infiltrate the ground within a day or two. This creates the advantage that the rain garden does not allow mosquitoes to breed.

History

The first rain gardens were created to mimic the natural water retention areas that occurred naturally before development of an area. The rain gardens for residential use were developed in 1990 in Prince George's County, Maryland, when Dick Brinker, a developer building a new housing subdivision had the idea to replace the traditional best management practices (BMP) pond with a bioretention area. He approached Larry Coffman, the county's Associate Director for Programs and Planning in the Department of Environmental Resources, with the idea. The result was the extensive use of rain gardens in Somerset, a residential subdivision which has a 300–400 sq ft (28–37 m^2) rain garden on each house's property. This system proved to be highly cost-effective. Instead of a system of curbs, sidewalks, and gutters, which would have cost nearly $400,000, the planted drainage swales cost $100,000 to install. This was also much more cost effective than building BMP ponds that could handle 2-, 10-, and 100-year storm events. Flow monitoring done in later years showed that the rain gardens have resulted in a 75–80% reduction in stormwater runoff during a regular rainfall event.

This is also referred to as Water Sensitive Urban Design (WSUD) in Australia, Sustainable Urban Drainage Systems or SUDS in the United Kingdom, and low impact development (LID) in the United States, and is cited by the EPA on their website as a success on the Stormwater Case Studies section of their website. This webpage has many links to information on Prince George's County's literature on implementing LID in a community.

Some *de facto* rain gardens predate their recognition by professionals as a significant LID tool. Any shallow garden depression implemented to capture and retain rain water within the garden so as to drain adjacent land without running off a property is at conception a rain garden — particularly if vegetation is maintained with recognition of its role in this function. Vegetated roadside swales, now promoted as "bioswales", remain the conventional drainage system in many parts of the world from long before extensive networks of concrete sewers became the conventional engineering practice in the industrialized world.

What is new about such technology is the emerging rigor of increasingly quantitative understanding of how such tools may make sustainable development possible. This is as true for wealthy developed communities retrofitting bioretention into built stormwater management systems, as for developing communities seeking a faster and more sustainable development path.

Characteristics

A home rain garden recently planted.

A rain garden requires an area where water can collect and infiltrate, and plants to maintain infiltration rates, diverse microbe communities, and water holding capacity. Transpiration by growing plants accelerates soil drying between storms. This includes any plant extending roots to the garden area.

Simply adjusting the landscape so that downspouts and paved surfaces drain into existing gardens may be all that is needed because the soil has been well loosened and plants are well established. However, many plants do not tolerate saturated roots for long and often more water runs off one's roof than people realize. Often the required location and storage capacity of the garden area must be determined first. Rain garden plants are then selected to match the situation, not the other way around.

Soil and Drainage

When an area's soils are not permeable enough to allow water to drain and filter properly, the soil should be replaced and an underdrain installed. This bioretention mixture should typically contain 60% sand, 20% compost, and 20% topsoil, and there is a current trend to replace compost with biochar. Existing soil must be removed and replaced. Do not combine the sandy soil (bioretention) mixture with a surrounding soil that does not have high sand content. Otherwise, the clay particles will settle in between the sand particles and form a concrete-like substance, as demonstrated in a 1983 study. Deep plant roots also create additional channels for storm water to filter into the ground. Microbial populations feed off plant root secretions and break down carbon (such as in mulch or desiccated plant roots) to aggregate soil particles which increases infiltration rates. A five-year USGS study indicates that rain gardens in urban clay soils can be effective without the use of underdrains or replacement of native soils with the bioretention mix. Pre-installation infiltration rates should be at least .25 in/

hour, however. Type D soils will require an underdrain paired with the sandy soil mix in order to drain properly.

Sometimes a drywell with a series of gravel layers near the lowest spot in the rain garden will help facilitate percolation. However, a drywell placed at the lowest spot can become clogged with silt prematurely turning the garden into an infiltration basin defeating its purpose. Depression-focused recharge of polluted water into wells poses a serious threat and should be avoided. Similarly plans to install a rain garden near a septic system should be reviewed by a qualified engineer. The more polluted the water, the longer it must be retained in the soil for purification. This is often achieved by installing several smaller rain garden basins with soil deeper than the seasonal high water table. In some cases lined bioretention cells with subsurface drainage are used to retain smaller amounts of water and filter larger amounts without letting water percolate as quickly.

Rain gardens are at times confused with bioswales. Swales slope to a destination, while rain gardens do not; however, a bioswale may end with a rain garden. Drainage ditches may be handled like bioswales and even include rain gardens in series, saving time and money on maintenance. Part of a garden that nearly always has standing water is a water garden, wetland, or pond, and not a rain garden. Using the proper terminology ensures that the proper methods are used to achieve the desired results.

Plant Selection

Plants selected for use in a rain garden should tolerate both saturated and dry soil. Using native plants is generally encouraged. This way the rain garden may contribute to urban habitats for native butterflies, birds, and beneficial insects.

Well planned plantings require minimal maintenance to survive, and are compatible with adjacent land use. Trees under power lines, or that up-heave sidewalks when soils become moist, or whose roots seek out and clog drainage tiles can cause expensive damage.

Trees generally contribute most when located close enough to tap moisture in the rain garden depression, yet do not excessively shade the garden. That said, shading open surface waters can reduce excessive heating of habitat. Plants tolerate inundation by warm water for less time because heat drives out dissolved oxygen, thus a plant tolerant of early spring flooding may not survive summer inundation.

Rain Garden Projects

Australia

- Healthy Waterways Raingardens Program promotes a simple and effective form of stormwater treatment, and aims to raise peoples' awareness about how good stormwater management contributes to healthy waterways. The program encourages people to build rain gardens at home, and has achieved its target is to see 10,000 rain gardens built across Melbourne by 2013.

- Melbourne Water's database of Water Sensitive Urban Design projects, including 57 case studies relating to rain gardens/bioretention systems. Melbourne Water is the Victorian State Government agency responsible for managing Melbourne's water supply catchments.
- Water By Design is a capacity building program that supports the uptake of Water Sensitive Urban Design, including rain gardens, in South East Queensland. It was established by the South East Queensland Healthy Waterways Partnership in 2005, as an integral component of the SEQ Healthy Waterways Strategy.

United Kingdom

- The Wildfowl and Wetlands Trust's London Wetland Centre includes a rain garden designed by Nigel Dunnett.
- Islington London Borough Council commissioned sustainable drainage consultants Robert Bray Associates to design a pilot rain garden in the Ashby Grove development which was completed in 2011. This raingarden is fed from a typical modest domestic roof catchment area of $30m^2$ and is designed to demonstrate how simple and cost effective domestic rain gardens are to install. Monitoring apparatus was built into the design to allow Middlesex University to monitor water volumes, water quality and soil moisture content. The rain garden basin is 300mm deep and has a storage capacity of $2.17m^3$ which is just over the volume required to store runoff from the roof catchment in a 1 in 100 storm plus 30% allowance for climate change.
- The Day Brook Rain Garden Project has introduced a number of rain gardens into an existing residential street in Sherwood, Nottingham.

United States of America

- The 12,000 rain garden campaign for Puget Sound is coordinating efforts to build 12,000 rain gardens in the Puget Sound Basin of Western Washington by 2016. The 12,000 rain gardens website provides information and resources for the general public, landscape professionals, municipal staff, and decision makers. By providing access to the best current guidance, easy-to-use materials, and a network of trained "Rain Garden Mentor" Master Gardeners, this campaign seeks to capture and cleanse over 200 Million gallons of polluted runoff each year, and thereby significantly improve Puget Sound's water quality.
- Maplewood, Minnesota has implemented a policy of encouraging residents to install rain gardens. Many neighborhoods had swales added to each property,

but installation of a garden at the swale was voluntary. The project was a partnership between the City of Maplewood, University of Minnesota Department of Landscape Architecture, and the Ramsey Washington Metro Watershed District. A focus group was held with residents and published so that other communities could use it as a resource when planning their own rain garden projects.

- In Seattle, a prototype project, used to develop a plan for the entire city, was constructed in 2003. Called *SEA Street,* for Street Edge Alternatives, it was a drastic facelift of a residential street. The street was changed from a typical linear path to a gentle curve, narrowed, with large rain gardens placed along most of the length of the street. The street has 11% less impervious surface than a regular street. There are 100 evergreen trees and 1100 shrubs along this 3-block stretch of road, and a 2-year study found that the amount of stormwater which leaves the street has been reduced by 99%.
- 10,000 Rain Gardens is a public initiative in the Kansas City, Missouri metro area. Property owners are encouraged to create rain gardens, with an eventual goal of 10,000 individual gardens.
- The West Michigan Environmental Action Council has established Rain Gardens of West Michigan as an outreach water quality program. Also in Michigan, the Southeastern Oakland County Water Authority has published a pamphlet to encourage residents to add a rain garden to their landscapes in order to improve the water quality in the Rouge River watershed. In Washtenaw County, homeowners can volunteer for the Water Resources Commissioner's Rain Garden program, in which volunteers are annually selected for free professional landscape design. The homeowners build the gardens themselves as well as pay for landscaping material. Photos of the gardens as well as design documents and drainage calculations are available online.
- The city of Portland, Oregon, has established a Clean River Rewards program, to encourage residents to disconnect downspouts from the city's combined sewer system and create rain gardens. Workshops, discounts on storm water bills, and web resources are offered.
- In Delaware, several rain gardens have been created through the work of the University of Delaware Water Resources Agency, and environmental organizations, such as the Appoquinimink River Association.
- In New Jersey, the Rutgers Cooperative Extension Water Resources Program has already installed over 125 demonstration rain gardens in suburban and urban areas. The Water Resources Program has begun to focus on using rain gardens as green infrastructure in urban areas, such as Camden and Newark to help prevent localized flooding, combined sewer overflows, and to improve water quality. The Water Resources Program has also revised and produced

a rain garden manual in collaboration with The Native Plant Society of New Jersey.

References

- Figures from the United Nations' *UN National Accounts Database.* for the countries of the world. Retrieved 1 February 2014. Johnson, Anthony, *Solving Stonehenge: The New Key to an Ancient Enigma.* (Thames & Hudson, 2008) ISBN 978-0-500-05155-9
- D. Masse; JL. Chotte; E. Scopel (2015) "Ecological engineering for sustainable agriculture in arid and semiarid West African regions". Fiche thématique du CSFD (11):2
- Howard R. Turner (1997), *Science in Medieval Islam: An Illustrated Introduction*, p. 181, University of Texas Press, ISBN 0-292-78149-0
- Davis, M. L. and D. A. Cornwell, (2006) *Introduction to environmental engineering* (4th ed.) McGraw-Hill ISBN 978-0072424119
- Compare: *"Construction", Merriam-Webster.com, Merriam-Webster, retrieved 2016-02-16, [...] the act or process of building something (such as a house or road) [...]*

Permissions

All chapters in this book are published with permission under the Creative Commons Attribution Share Alike License or equivalent. Every chapter published in this book has been scrutinized by our experts. Their significance has been extensively debated. The topics covered herein carry significant information for a comprehensive understanding. They may even be implemented as practical applications or may be referred to as a beginning point for further studies.

We would like to thank the editorial team for lending their expertise to make the book truly unique. They have played a crucial role in the development of this book. Without their invaluable contributions this book wouldn't have been possible. They have made vital efforts to compile up to date information on the varied aspects of this subject to make this book a valuable addition to the collection of many professionals and students.

This book was conceptualized with the vision of imparting up-to-date and integrated information in this field. To ensure the same, a matchless editorial board was set up. Every individual on the board went through rigorous rounds of assessment to prove their worth. After which they invested a large part of their time researching and compiling the most relevant data for our readers.

The editorial board has been involved in producing this book since its inception. They have spent rigorous hours researching and exploring the diverse topics which have resulted in the successful publishing of this book. They have passed on their knowledge of decades through this book. To expedite this challenging task, the publisher supported the team at every step. A small team of assistant editors was also appointed to further simplify the editing procedure and attain best results for the readers.

Apart from the editorial board, the designing team has also invested a significant amount of their time in understanding the subject and creating the most relevant covers. They scrutinized every image to scout for the most suitable representation of the subject and create an appropriate cover for the book.

The publishing team has been an ardent support to the editorial, designing and production team. Their endless efforts to recruit the best for this project, has resulted in the accomplishment of this book. They are a veteran in the field of academics and their pool of knowledge is as vast as their experience in printing. Their expertise and guidance has proved useful at every step. Their uncompromising quality standards have made this book an exceptional effort. Their encouragement from time to time has been an inspiration for everyone.

The publisher and the editorial board hope that this book will prove to be a valuable piece of knowledge for students, practitioners and scholars across the globe.

Index

www.ingramcontent.com/pod-product-compliance
Lightning Source LLC
LaVergne TN
LVHW080723210926
848485LV00003B/511